# Upcycling Legume Water: from wastewater to food ingredients

Luca Serventi

# Upcycling Legume Water: from wastewater to food ingredients

Luca Serventi
Faculty of Agriculture and Life Sciences
Lincoln University
Christchurch, New Zealand

ISBN 978-3-030-42470-1 ISBN 978-3-030-42468-8 (eBook)
https://doi.org/10.1007/978-3-030-42468-8



This Springer imprint is published by the registered company Springer Nature Switzerland AG
The registered company address is: Gewerbestrasse 11, 6330 Cham, Switzerland

*To my love, Lindy, I dedicate this book*

# Foreword

Food manufacturing generates wastewater. The legume industry contributes to this environmental issue with the processing of legume seeds into food products. Current technologies result in nutrient loss and environmental damage.

Therefore, upcycling legume wastewater should become a common practice in the food industry of the future. Recycling by-products into value-added food ingredients can maintain nutrients in the food chain and reduce the load on the environment.

From waste to food ingredients, the proposed innovation is to reduce waste, to optimize the use of natural resources and to add value to product development. New exciting opportunities await the future generations of food innovators.

Lincoln University
Lincoln, New Zealand

Luca Serventi,

# Preface

Soaking, boiling and sprouting are processes needed to transform dried legumes into foods and food products such as soymilk, tofu, hummus and sprouts. Large volumes of liquid by-products are generated and must be treated prior to dismissal into the environment, thus raising production costs for the food industry. In addition, wastewater contains nutrients that are lost from the food chain. Water and soluble nutrients are becoming scarce due to an increasing world population, and it is critical to optimize food manufacturing.

An alternative to waste treatment is upcycling, a concept of circular economy. Circular economy proposes to reuse all resources involved in manufacturing by recycling and/or upcycling. Upcycling refers to the process of finding a new use for by-products, with higher value than the raw material. Legume wastewater includes the liquids discarded from soaking, cooking and sprouting of legumes, and, in this book, it is called "Liluva". Liluva used to be an environmental concern, but in the context of circular economy, it is an exciting opportunity for sustainable food production.

Upcycling Liluva in food ingredients is proposed in this book. Chapters 1 and 2 introduce legume composition and functionality, as well as the generation of wastewater during their processing. Composition, functionality and applications are described for soaking water (Chaps. 3, 4 and 5) and cooking water (Chaps. 6, 7 and 8). Furthermore, a new concept is introduced in Chap. 9: sprouting water. Also, nutritional evaluation of legume bioactives is included in Chap. 10. Finally, Liluva is discussed as a source of edible packaging in Chap. 11. Applications in food products consist of, but are not limited to, emulsifiers, foaming agents, thickeners, egg replacers, nutritional supplements, prebiotics and enzymes. Liluva can replace expensive hydrocolloids as functional ingredients (e.g. in gluten-free products) and to extend shelf-life (antistaling agents). In addition, it enhances the nutritional value of bakery, confectionery, spreads and other food products, delivering fibre, protein, minerals, phytochemicals and prebiotics. Finally, it is a source of enzymes for textural improvement of high-protein and high-fibre foods. Applications in edible packaging have been considered as well. New experimental results are presented for the first time in Chaps. 3, 5, 6, 7, 8 and 9.

To the best of my knowledge, this book is a world first because it is the result of teaching. Data collection and writing have been achieved through research essays and research placements. The contributors, at the time of their work, are six bachelor students, four postgraduate diploma students, seven taught master students and two faculties (including me). Now, they are off to industry jobs and further studies.

This book is thought for innovators and product developers in the industries of food, food ingredients, nutritional supplements and bioplastics. It is also directed to schools and universities to teach recycling and upcycling of food, with a focus on legume water (Liluva). I hope that you enjoy this book, be inspired by it and pass on that spark of optimistic innovation to the next generations.

Lincoln, New Zealand Luca Serventi

# Acknowledgements

This book represents my passion for food science, innovation and exploring the world. In 2015, these motivations took me to a small town of New Zealand, Lincoln, to teach at a local university. Here, I met people who changed my life and showed me the way. Four years later, the discoveries made with students, colleagues and friends are staggering. It's about time to share them with the world, academic and industrial, to impact how we make food. I believe that upcycling Liluva (legume wastewater) is a great way to promote sustainability and health and to improve our lives.

All of this was possible because of the unconditioned support of my parents, best friends and love. Thank you mom and dad, for encouraging my journey. Thank you Thomas Corradi, Ben Yeap, Bryan Finfrock, Michael Finfrock, Missy Utz-Finfrock, Dr Venkata Chelikani and Dr Federico Tomasetto, for being the wise, reliable friends who give me certainties. Thank you Lindy, for your love.

To all of you I say: Grazie di cuore!

# Contents

# Contributors

**Yaqi Bian** Department of Wine, Food and Molecular Biosciences, Faculty of Agriculture and Life Sciences, Lincoln University, Lincoln, Christchurch, New Zealand

**Yuxin Cai** Department of Wine, Food and Molecular Biosciences, Faculty of Agriculture and Life Sciences, Lincoln University, Lincoln, Christchurch, New Zealand

**Wendian Chang** Department of Wine, Food and Molecular Biosciences, Faculty of Agriculture and Life Sciences, Lincoln University, Lincoln, Christchurch, New Zealand

**Venkata Chelikani** Department of Wine, Food and Molecular Biosciences, Faculty of Agriculture and Life Sciences, Lincoln University, Lincoln, Christchurch, New Zealand

**Mingyu Chen** Department of Wine, Food and Molecular Biosciences, Faculty of Agriculture and Life Sciences, Lincoln University, Lincoln, Christchurch, New Zealand

**Hoi Tung Chiu** Department of Wine, Food and Molecular Biosciences, Faculty of Agriculture and Life Sciences, Lincoln University, Lincoln, Christchurch, New Zealand

**Sachin Deshmukh** Department of Wine, Food and Molecular Biosciences, Faculty of Agriculture and Life Sciences, Lincoln University, Lincoln, Christchurch, New Zealand

**Lirisha Vinola Dsouza** Department of Wine, Food and Molecular Biosciences, Faculty of Agriculture and Life Sciences, Lincoln University, Lincoln, Christchurch, New Zealand

**Congyi Gao** Department of Wine, Food and Molecular Biosciences, Faculty of Agriculture and Life Sciences, Lincoln University, Lincoln, Christchurch, New Zealand

**Paramjot Kaur** Department of Wine, Food and Molecular Biosciences, Faculty of Agriculture and Life Sciences, Lincoln University, Lincoln, Christchurch, New Zealand

**Jiaying Lin** Department of Wine, Food and Molecular Biosciences, Faculty of Agriculture and Life Sciences, Lincoln University, Lincoln, Christchurch, New Zealand

**Silu Liu** Department of Wine, Food and Molecular Biosciences, Faculty of Agriculture and Life Sciences, Lincoln University, Lincoln, Christchurch, New Zealand

**Yaying Luo** Department of Wine, Food and Molecular Biosciences, Faculty of Agriculture and Life Sciences, Lincoln University, Lincoln, Christchurch, New Zealand

**Neha Nair** Department of Wine, Food and Molecular Biosciences, Faculty of Agriculture and Life Sciences, Lincoln University, Lincoln, Christchurch, New Zealand

**Luca Serventi** Department of Wine, Food and Molecular Biosciences, Faculty of Agriculture and Life Sciences, Lincoln University, Lincoln, Christchurch, New Zealand

**Dan Xiong** Department of Wine, Food and Molecular Biosciences, Faculty of Agriculture and Life Sciences, Lincoln University, Lincoln, Christchurch, New Zealand

**Yiding Yang** Department of Wine, Food and Molecular Biosciences, Faculty of Agriculture and Life Sciences, Lincoln University, Lincoln, Christchurch, New Zealand

**Yanyu Zhang** Department of Wine, Food and Molecular Biosciences, Faculty of Agriculture and Life Sciences, Lincoln University, Lincoln, Christchurch, New Zealand

**Jingnan Zhu** Department of Wine, Food and Molecular Biosciences, Faculty of Agriculture and Life Sciences, Lincoln University, Lincoln, Christchurch, New Zealand

# Chapter 1
# Introduction: Legume Processing

**Paramjot Kaur and Luca Serventi**

## 1.1 Legumes and Diet

Human nutrition relies on dry grains from nearly 20 leguminous species. Common beans (*Phaseolus vulgaris L.*), chickpeas (*Cicer arietinum L.*) and peas (*Pisum sativum L.*) are used in many diets, including Asian and African countries (de Almeida de Almedia Costa et al. 2006). Legumes are a source of carbohydrates, protein, dietary fibre, lipids, minerals and vitamins (Nielsen 1991). They belong to family *Leguminosae*, which are dicotyledonous seeds, and consist of more than 17,000 species and 600 genera (Du et al. 2014). The main nutrients are carbohydrates (starch and fibre) and protein (Fig. 1.1). Carbohydrate content ranges from 30 to 75 g/100 g for soy and lima beans, respectively, while protein content spans from 15 to 41 g/100 g for pigeon pea and soy, respectively (USDA 2019). Grain legumes such as chickpeas, lentils and peas are rich source of dietary fibre and protein, higher than grains consumed more often such as rice, corn and wheat. In contrast, low fat content is present in legumes with the only exception of peanuts and soy (Baik & Han 2012). The amount of minerals in legumes is sufficient to fulfil the mineral requirement in humans (Iqbal et al. 2006) including proteins of high biological value. Legumes are rich in lysine which is one of essential amino acid, but lacking sulphur-containing amino acids such as cystine, cysteine and methionine (Iqbal et al. 2006).

Generally, legumes are consumed after processing. Cooking of legumes increases the bioavailability of nutrients and sensory quality. It also inactivates trypsin, haemagglutinins and growth inhibitors (Tharanathan and Mahadevamma 2003). Cereals contain more digestible carbohydrate than legumes. Carbohydrate helps in

P. Kaur · L. Serventi (✉)
Department of Wine, Food and Molecular Biosciences, Faculty of Agriculture and Life Sciences, Lincoln University, Lincoln, Christchurch, New Zealand
e-mail: Luca.Serventi@lincoln.ac.nz



L. Serventi, *Upcycling Legume Water: from wastewater to food ingredients*,
https://doi.org/10.1007/978-3-030-42468-8_1

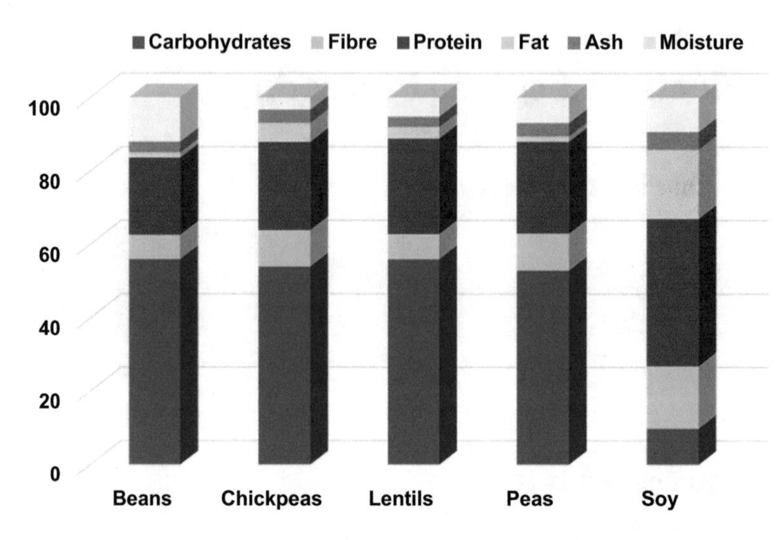


**Fig. 1.1** Representative proximate compositions of five commonly eaten legumes in g/100 g (de Almeida Costa et al. 2006; Iqbal et al. 2006; Meiners et al. 1976; Redondo-Cuenca et al. 2007)

the control of postprandial blood glucose and thereby reduces the risk of type II diabetes (Baik and Han 2012). Raw legumes have less insoluble dietary fibre than freeze dried cooked legumes. Physicochemical and nutritional properties of legumes can be reduced by cooking whereas the nutrient content increases by freeze drying (de Almeida Costa et al. 2006). A decrease in naturally existing antinutritional factors is observed in legumes by thermal treatment which results in increased availability of starch and protein (de Almeida Costa et al. 2006).

Legumes are beneficial for treating cancer and cardiovascular diseases as they are rich in bioactive phytochemicals. Apart from this, low starch digestibility and high fibre content of legumes helps in treating health problems such as obesity, diabetes and high blood cholesterol and preventing endemic diseases (Baik and Han 2012). Legumes are rich source of phytochemicals, which include antioxidants, phytosterols and bioactive carbohydrates. These phytochemicals can reduce tumour risk when consumed in sufficient amount (Kalogeropoulos et al. 2010). Among legumes, chickpea is most consumed across the world. It consists of two parts, outer part of seed coat and inner part of cotyledons. In other words, granules of chickpea are rich in starch which is embedded in protein coat. These legumes are cooked to improve their nutritional content in their seeds so that they are good in taste as well. It is common process of soaking of legumes before cooking. This process is practiced mostly so that water can be absorbed and distributed between seed and outer protein layer and less time is consumed for cooking (Chenoll et al. 2009).

Dried legumes are hard in their appearance as such cooking and soaking in water enables them to appear soft in texture. Soaking and cooking affect texture of legumes as they degrade some unimportant substances which may affect digestion (Anzaldua-Morales et al. 1996). In an experiment conducted by Chenoll and associates (2009), results revealed that starch reduced by almost 20% in cooked chickpea in comparison to raw chickpea. One of the possible reasons is the solubilisation of carbohydrates (in particular α-galactoside) in the cooking water. Seed size increases after 120 minutes of soaking and outer layer of chickpea does not as act as barrier. Anzaldua and associates (1996) studied the effect of cooking on softening of different legumes and concluded that raw chickpea is hardest in appearance yet cooking time was much less to other legumes. Starch gelatinization occurs in chickpea upon cooking which results in swelling of seed. Chickpea acted as base when mixed with other batters and dough. A research done by Domene and Oliveira (1993) showed that by thermal treatment of legumes (as cooking), naturally occurring antinutritional factors decreased and the availability of other nutrients such as protein and starch increases thereby making the consumption of legumes more convenient.

Research findings have shown that the successful performance of legume flour depends on the functional characteristics of legumes which they impart to final food product. Properties such as emulsification, foaming, water and oil absorption capacities, viscosity and gelation form a part of functional characteristics of legume (Adebowale and Lawal 2004). A report by Du and collaborators (2014) stated that bulk density varies among different legume flours. Lentil flour was denser than those of other legumes as it had higher bulk density. Starch molecules swell as the volume increases in presence of excess water which is indicated in water absorption index along with integrity of starch molecules in aqueous dispersions. Similarly, solubility of molecules is indicated in water soluble index and it varies in different legume flours. Functional and sensory properties of foods are influenced by water absorption capacity of the flour used. Oil absorption capacity of legume flours greatly affects mouthfeel and flavour of food products. Capillary action is involved in oil absorbing mechanism which helps in retaining of absorbed oil in food and protein have important role in it (Du et al. 2014). Poor emulsion activity is indicated in chickpea flour as it has lower protein content than other legume flours. On the contrary, higher protein content was observed in lentil flours but with poor emulsion activity. According to Baik and Han (2012), oligosaccharide content of legumes decreases on cooking and roasting in comparison to raw legumes. Roasted and cooked legumes have better protein digestibility with lower protein solubility. Higher content of starch is present in cooked and roasted chickpeas, peas and lentils as compared to raw legumes. During cooking, soluble components of legumes were lost, which resulted in increased starch content of cooked legumes. Raw form of chickpeas, lentils have higher protein content than cooked legumes.

Functionality of a food product is defined by the properties of its ingredients that have impact on its utilization. Properties such as viscosity, swelling, water absorption, properties related to hydration and binding have impact on characteristics of food. Acceptance of protein depends greatly on functional properties in food systems (Kaur and Singh 2005). Therefore, the objective of this chapter is to discuss the nutritional content, physiochemical properties and applications of legume ingredients in food processing.

## 1.2 Nutrients in Legumes

### *1.2.1 Starch*

Starch is the main source of energy in the human diet. It is commonly known as available carbohydrate. Starch is present in different forms in processed foods: partially changed, wholly changed and not retrograded. Gelatinized starches is hydrolysed by enzymes whereas raw starches are unaffected by enzymatic hydrolysis. Hydrolysis is enhanced by starch gelatinization. During gelatinization, starch undergoes phase transition upon heating in presence of water which is irreversible. The increase in granules size can be related to disruption of semi crystalline structure of starch. Insulin responses and post prandial glucose responses are stimulated by legumes both moderately and slowly as a result of poor starch digestibility. Legumes are considered beneficial in hyperlipidaemia and diabetes as legumes are considered as sloe release carbohydrate (Tharanathan and Mahadevamma 2003). The level of amylose lipid complexes and amylose differ in starches of legumes. Even starch chain associations vary in native granule (Hoover and Manuel 1996). In legumes, starch is one of the major biopolymeric constituent. Upon processing, it is partially modified into resistant starch. Amylose, which is of high molecular weight and branched, contributes to reduced and slow digestibility of starch in legumes (Tharanathan and Mahadevamma 2003). Legumes contain approximately 50 g/100 g starch, 35% of which being amylose (Du et al. 2016). The higher the content of amylose, the higher is resistant starch which imparts its effect as low digestibility. Decreased glycemic responses in legumes are also result of the protein-starch interaction as legumes contain more protein than cereals (Hoover and Manuel 1996).

### *1.2.2 Dietary Fibre*

Dietary fibre can be defined as polysaccharides and lignin that are not digested in the human digestive system by endogenous secretions inside the human body and it is an important component of foods (Azizah and Zainon 1997). Mainly, fibre is present in two forms: soluble dietary fibre (SDF) and insoluble dietary fibre (IDF). Raw legumes have less amount of SDF in comparison to freeze dried cooked legumes. Freeze drying and thermal treatment mark minor rise in the amount of nutrients (de Almeida Costa et al. 2006).

Dietary fibre in legumes contributes to health benefits as it is a heterogeneous mixture of a number of different polysaccharides, present in husks as well. The antinutritional factors amylase and phytates inhibitors and high amount of dietary fibre affect digestibility of legume starch greatly (Tharanathan and Mahadevamma 2003). With thermal treatment, both SDF and IDF increases in samples with high protein content which can be related to production of Maillard's reaction products (Azizah and Zainon 1997). The TDF and IDF are the highest in cooked chickpea and SDF is highest in raw chickpea, while insoluble dietary fibre and total dietary fibre of boiled chickpea are significantly less in comparison to raw chickpea values (Perez-Hidalgo et al. 1997). It has been proposed that resistant starch formation and Maillard's reaction products along with condensed tannin protein products may lead to dietary fibre increase as dry matter is lost in cooking (Perez-Hidalgo et al. 1997). Another study conducted on boiled legumes showed low fibre content for soybeans (0.9 g/90 g) and higher values for kidney beans, chickpeas and lentils (3.2, 2.9 and 4.0 g/90 g, respectively (Messina 1999).

### *1.2.3 Protein*

Pulse legumes contain protein in amounts ranging from 22 to 28 g/100 g of chickpeas and lentils, respectively (Du et al. 2016; Iqbal et al. 2006). Protein content and digestibility differ in legumes based on cultivar and processing. In chickpea, protein digestibility is increased by roasting and cooking, more substantially than for peas and lentils (Baik and Han 2012). Protein digestibility is comparable among lentils, soybean and chickpea in fermented legumes but is lower than in raw legumes (Baik and Han 2012). The reason behind low protein digestibility in fermented legumes with respect to roasted and cooked ones is the reduction of pH due to hydrolysis of protein by proteases excreted by molds in the process of fermentation. The hydrolysis of protein result in release of amino acids by activity of different pH medium which is beneficial in assessment of digestibility of protein and digestibility value (Baik and Han 2012). In cooked, roasted and fermented legumes, lesser digestibility of protein is observed except that for cooked peas (Baik and Han 2012). Protein denaturation is caused by heating in legumes which further result in heat induced aggregates, possibly contributes to loss of solubility (Carbonaro et al. 1997). Similarly, protein structure is altered by addition of sodium chloride which results in destabilization of emulsions (Makri et al. 2005). In chickpea, sulphur-containing amino acids are the most limiting ones whereas tryptophan is scarce in lentil, cowpea and green pea (Iqbal et al. 2006).

Albumin protein exhibits foaming ability and it is present mainly in peas and chickpeas (14 and 12 g/100 g, respectively) while lower amount was found in lentils (8.1 g/100 g) (Iqbal et al. 2006). Roasting causes maximum reduction in protein digestibility in peas, beans and chickpeas. Roasting reduces protein solubility by 21%, 35%, 37% and 22% in lentils, chickpeas, beans and peas respectively. In cooked, roasted and fermented lentils protein solubility is less than 11% and raw lentils contain 29% of protein solubility. Pressure cooking for 50 minutes reduces protein solubility in chickpeas from 43% to 4%. Proteins are hydrolyzed during fermentation. The protein solubility reduces from fermented, cooked to roasted beans as 25%, 8.5% and 1.5% respectively and higher protein solubility in fermentation can be result of high amount of protein hydrolysis (Baik and Han 2012). Legume proteins have functional and nutritional properties which contribute their vital role in formulation and processing of food (Miñarro et al. 2012). In legume seeds, albumin protein (enzymatic and non-storage protein) is present in small amount while globulin protein (storage) is considered as a major fraction. The albumin protein of chickpeas and peas is rich in sulphur-containing amino acid and lysine as compared to globulin protein. Peas contained higher levels of albumin protein than chickpeas and peas, the highest amount among legumes (Bhatty 1982).

Mostly, globulins are storage protein in grain of legumes. Seeds of legume grains contain antinutritional compounds, proteinous in nature such as lectins and hydrolase inhibitors (Duranti 2006). The quality of legume seeds is determined by seed hydrolase inhibitors. Usually, heat denatures these proteins and their effect is only observed in uncooked seeds and flour (Duranti and Gius 1997). Insolubility of protein is reduced by denaturation of protein during roasting and cooking. Heat and moisture is provided to legumes through cooking. Dry heat is applied by roasting which is insufficient for starch gelatinization in legumes. More starch is gelatinized in cooked legumes than roasted legumes. The water absorption capacity is more in ungelatinized starch than in gelatinized starch. In comparison to wheat flour, legume flour has high content of soluble and insoluble fiber and soluble protein (Baik and Han 2012).

### *1.2.4 Lipids*

Different amounts of polyunsaturated fatty acids are present in legume such as, linoleic acid (18:2), representing 21% of lipids in peas (Grela and Günter 1995). Oleic acid was the highest in chickpeas, followed by peas and lentils. Chickpeas contain the highest amount of linoleic acid and alpha-linolenic acid (ALA), both essential fatty acids (Caprioli et al. 2016; Grela and Günter 1995).

Lipids affect food texture. For example, in bakery products, gas vacuoles form small holes in the starch protein matrix of the dough. In foam structure, gas vacuoles appear as small cells dispersed in matrix. During fermentation, these small vacuole, undergo rupture, thus releasing carbon dioxide in the dough as a result of gelatinisation of starch which increases viscosity. The presence of polar lipids in legume flour helps in retention of gas vacuole by forming a lipid monolayer in starch protein matrix and gas interface (Gan et al. 1995). In an experiment conducted by Gan and associates (1995), bread loaf volume increased with presence of polar lipids in legume flours. In contrast, non-polar lipids in loaf decreased volume of bread. As such it is clear besides other beneficial effects of legume flours, it can be utilised in bread giving them good volume.

### 1.2.5 Micronutrients and Antinutrients

Chickpeas, peas and lentils contain minerals in different proportions, with the most abundant being potassium (around 100 mg/100 g) and calcium (110–197 mg/100 g), as well as relevant amounts of iron (3–10 mg/100 g in chickpeas and peas, respectively) (Iqbal et al. 2006).

A number of anti-nutritional compounds are present in legumes (Duranti and Gius 1997). Alkaloids confer legumes an unacceptable taste. Phenolic compounds can react with residues of methionine and lysine and make less availability of protein during digestion. Saponins are responsible for limited absorption of protein, sugar and cholesterol (Duranti and Gius 1997). Sometimes, flatulence is caused by legumes. The reason is linkage of alpha-D-galactopyranosyl (absent in humans) with sucrose. In humans, in colon alpha-galactosidase are broken down by intestinal flora which results in production of gas (Duranti and Gius 1997).

## 1.3 Legumes as Food Ingredients (Functional Properties)

### 1.3.1 Legume Flours

Legumes can be ground into flour and used as food ingredient due to their high protein content. Physical and chemical properties such as structure, hydrophobicity, amino acid sequence and composition, molecular weight and structure are closely associated with the functionality of proteins, along with presence of salts, sugars, proteins, water and fat. The functional characteristics of food ingredients include foaming, emulsifying activity index (EAI), as well as water and oil absorption capacities (Table 1.1). The effective enactment of legume flour as a food ingredient depends on the above mentioned functional characteristics (Du et al. 2014). Differences in genetic information, growth and environmental conditions of legumes

**Table 1.1** Physicochemical properties (foaming, emulsifying and thickening) of legume flours

| Legume flour | FA (%) | EAI ($m^2/g$) | WAC (g/g) | OAC (g/g) | References |
|---|---|---|---|---|---|
| Navy beans | 80–92 | 13 | 1.4–2.6 | 1.2–2.2 | Du et al. (2014), Gupta et al. (2018) and Wani et al. (2013) |
| Chickpeas | 50–175 | 10 | 1.6–6.1 | 1.1 | Du et al. (2014) and Gupta et al. (2018) |
| Lentils | 80–175 | 14 | 1.8–4.8 | 0.9 | Du et al. (2014) and Gupta et al. (2018) |
| Peas | 40–175 | 13 | 1.3 | 1.1–1.2 | Maninder et al. (2007) |
| Soybeans | 0.81–0.88[a] | 11–18 | 3.1–6.7 | 1.4–2.2 | Gupta et al. (2018) and Heywood et al. (2002) |

[a]mL foam/mL $N_2$/min

result in variation of chemical composition of different flours. In food formulations, bulk density plays an important role. For example, lentil flour is denser in comparison to other legumes because of a higher bulk density (Du et al. 2014). In addition, water absorption capacity of legume flours alters the texture of finished food products. For instance, addition of legume flour to baking formula results in softer texture. The integrity of starch molecules is indicated by water absorption index (WAI) in aqueous solution and it is higher in chickpea flour (Du et al. 2014). The cooking properties of flours from legumes are correlated with water absorption capacity (WAC). In order to maintain flavour and improve mouthfeel, the oil absorbing capacity (OAC) plays an important role, being affected by starch and protein content, the type of protein, size of particles and non-polar amino acid side chain ratios of the surface of protein molecule in the legume flour. Furthermore, hydrophobic proteins play an important role in oil absorption. Similarly, capillarity interaction is involved in oil absorbed mechanism (Du et al. 2014). Different legume flours have different emulsion activity, with chickpea exhibiting the lowest emulsion activity. The reason behind this is lower protein content in comparison to other legume flours. In contrast, the protein content of lentil flour is the highest among pulse legumes but its emulsion activity is also poor. Such variation can be related to the other components in the flours such as starch and fat content. Chickpea has the lowest foam stability, foaming capacity and lowest fat content among legumes (Du et al. 2014).

### *1.3.2 Cooked Legume Flours*

Cooked legumes deteriorate quickly, thus further impacting the process of digestion (Tharanathan and Mahadevamma 2003). Higher content of starch is present in cooked and roasted lentils, peas and chickpeas in comparison to raw legumes (Baik and Han 2012). Fermented peas and chickpeas have decreased content of starch. Loss of soluble components of legumes during cooking probably contributes to the relative increase in starch content of cooked legumes (Baik and Han 2012).

During fermentation, starch hydrolytic enzymes are released by the mould, which decreases the starch content of fermented legumes. Higher protein content in cooked and fermented chickpea, soybean and lentils is present in comparison to raw legumes. The protein content in cooked and fermented soybeans is 3–7% higher than in raw soybeans. Protein content increases in cooked lentils and chickpeas by approximately 1%. Soaking and cooking decrease protein content in dried legumes and reduction can be related to drain off protein content to cooking water. In cooked legumes, protein content increase and decrease is related to presence of other soluble components and their comparison with drained off content of protein during process of cooking (Baik and Han 2012).

A decrease in oligosaccharide content in legumes is observed in processes such as fermentation, roasting and cooking and this decrease is 28% in peas, 43% in soybeans and 34% in chickpeas. Smaller seed size of lentils can be cause of grater reduction in lentil oligosaccharide content with respect to other legumes (Baik and

Han 2012). In fermented legumes, the content of oligosaccharide is as low as 3% of that content is present in comparison to raw chickpeas. The same content reduces to 24% in peas (Baik and Han 2012). In peas, soybeans and chickpeas roasting result in greater reduction in content of oligosaccharide in comparison to cooking and reverse is observed in lentils, where cooking is active than roasting. The content of oligosaccharide present in roasted legume is almost 40 times that of raw legumes (Baik and Han 2012). Starch is highest in cooked chickpea flour than roasted and cooked chickpea flour. Protein is highest in fermented chickpea flour and least in raw chickpea flour. For lentils highest starch content is present cooked form and least in raw lentils. A small change in protein percentage is observed in raw, cooked, fermented and roasted lentils. For peas, starch is more available in cooked and roasted flour in comparison to raw and fermented flour. Protein is highest in fermented and roasted pea flour than from raw pea flour and least in cooked pea flour. The pasting properties differ in various legumes and this value is lowest for chickpea. Bean flour has highest value which further specifies that the starch present in it is highly unaffected by rupture and swelling. Following are properties of chickpea and lentil flour (Baik and Han 2012).

### *1.3.3 Food (Bread) Quality*

In an experiment conducted by Baik and Han (2012), the effect of legume flour on baking performance of a composite wheat-legume (70:30) bread was investigated. Mixograph absorption and loaf volume of bread were observed. An increase in Mixograph absorption from 68% to 76% was observed for cooked legume flours and from 62% to 64% for roasted legume flour. A possible reason behind this is denaturation of protein and gelatinization of starch in process of cooking and roasting. The effect of cooking is higher than that roasting as both heat and moisture were applied in cooking, whereas dry heat only was applied by roasting. It was observed that sticky and weak dough was produced from wheat and raw legume in comparison to whole wheat dough. Cooking and roasting improved gluten development by reducing stickiness. Bigger loaf volume of bread was produced from chickpea flour and wheat flour in comparison to other legume flours of pea and lentils. It was found that higher starch content is present in cooked and roasted chickpea, lentil and peas in comparison to raw ones. The explanation could be the loss of soluble content during cooking into cooking water. Similarly, cooked chickpeas, lentils and peas contained more protein than their raw counterparts, as a consequence of loss of soluble fibre, sugar and minerals into cooking water during cooking.

Proteins are denatured and gelatinization of starch occurs upon cooking. In comparison to wheat flour, legume flours are rich in protein along with soluble and insoluble fibre. The dough produced by mixing legume and wheat flour was sticky and weak in comparison to whole wheat flour dough. The reason behind this is the presence of protein and starch components in hydrolysed form in fermented legumes and the development of gluten-networks in cooked and roasted legumes. Roasted legume dough gave better results than raw and cooked legume dough. A bigger loaf of bread can be produced from chickpea and wheat flour as compared with pea and lentil flour. As such it is concluded that cooking, roasting has effect on the composition of legumes as compared to raw form of legumes. Apart from this, the functional properties of legumes get altered. Such are seen in volume of bread produced from different legumes flours in different processed forms (Baik and Han 2012). In an experiment conducted by Miñarro and associates (2012), it was observed that chickpeas expressed higher foam expansion than peas.

## 1.4 Conclusions

Legumes are rich in starch and proteins and can be used in various food formulations such as bakery products. Chickpeas contain high levels of starch and are rich in albumin, globulin, soluble dietary fibre and polyunsaturated fatty acids, although their protein content is lower than other legumes, with soy being the highest protein source (about 40 g/100 g). Insoluble dietary fibre is more present in lentils than other legumes. Legume flours in raw and cooked form were analysed for their nutritional content and it was concluded that starch is the main nutrient in roasted lentils, while protein are maximum in roasted peas. Legume flours exhibited emulsion activity mainly due to protein. These grains can be used for gluten-free breads as well as for nutraceutical application. They can be utilised as whole or fractions in several food products.

**Acknowledgments** This book chapter was written mainly by a Graduate Diploma student in partial completion of the course named "FOOD 398 – Research Essay" offered by Lincoln University.

## References

Adebowale, K. O., & Lawal, O. S. (2004). Comparative study of the functional properties of bambarra groundnut (*Voandzeia subterranean*), jack bean (*Canavalia ensiformis*) and mucuna bean (*Mucuna pruriens*) flours. *Food Research International, 37*(2004), 355–365.

Anzaldua-Morales, A., Quintero, A., & Balandran, R. (1996). Kinetics of thermal softening of six legumes during cooking. *Journal of Food Science, 61*(1), 167–170.

Azizah, A. H., & Zainon, H. (1997). Effect of processing on dietary fiber contents of selected legumes and cereals. *Malaysian Journal of Nutrition, 3*(1), 131–136.

Baik, B. K., & Han, I. H. (2012). Cooking, roasting, and fermentation of chickpeas, lentils, peas, and soybeans for fortification of leavened bread. *CerealChemistry, 89*(6), 269–275.

Bhatty, R. S. (1982). Albumin proteins of eight edible grain legume species. Electrophoretic patterns and amino acid composition. *Journal ofAgricultural and Food Chemistry, 30*(3), 620–622.

Caprioli, G., Giusti, F., Ballini, R., Sagratini, G., Vila-Donat, P., Vittori, S., & Fiorini, D. (2016). Lipid nutritional value of legumes: Evaluation of different extraction methods and determination of fatty acid composition. *Food Chemistry, 192*, 965–971.

Carbonaro, M., Cappelloni, M., Nicoli, S., Lucarini, M., & Carnovale, E. (1997). Solubility-digestibility relationship of legume proteins. *Journal of Agricultural and Food Chemistry, 45*(9), 3387–3394.

Chenoll, C., Betoret, N., & Fito, P. (2009). Analysis of chickpea (var."Blanco Lechoso") rehydration. Part I. Physicochemical and texture analysis. *Journal of Food Engineering, 95*(2), 352–358.

de Almeida Costa, G. E., da Silva Queiroz-Monici, K., Reis, S. M. P. M., & de Oliveira, A. C. (2006). Chemical composition, dietary fibre and resistant starch contents of raw and cooked pea, common bean, chickpea and lentil legumes. *Food Chemistry, 94*(3), 327–330.

Domene, S. M. A., & Oliveira, A. C. (1993). The use of nitrogen-15 labeling for the assessment of leguminous protein digestibility. *Journal of Nutritional Science and Vitaminology, 39*(1), 47–53.

Du, S. K., Jiang, H., Yu, X., & Jane, J. L. (2014). Physicochemical and functional properties of whole legume flour. *LWT-Food Science and Technology, 55*(1), 308–313.

Duranti, M. (2006). Grain legume proteins and nutraceutical properties. *Fitoterapia, 77*(2), 67–82.

Duranti, M., & Gius, C. (1997). Legume seeds: Protein content and nutritional value. *Field Crops Research, 53*(1), 31–45.

Gan, Z., Ellis, P. R., & Schofield, J. D. (1995). Gas cell stabilisation and gas retention in wheat bread dough. *Journal of Cereal Science, 21*(3), 215–230.

Grela, E. R., & Günter, K. D. (1995). Fatty acid composition and tocopherol content of some legume seeds. *Animal Feed Science and Technology, 52*(3), 325–331.

Gupta, S., Chhabra, G. S., Liu, C., Bakshi, J. S., & Sathe, S. K. (2018). Functional properties of select dry bean seeds and flours. *Journal of Food Science, 83*(8), 2052–2061.

Heywood, A. A., Myers, D. J., Bailey, T. B., & Johnson, L. A. (2002). Functional properties of extruded-expelled soybean flours from value-enhanced soybeans. *Journal of the American Oil Chemists' Society, 79*(7), 699–702.

Hoover, R., & Manuel, H. (1996). Effect of heat—moisture treatment on the structure and physicochemical properties of legume starches. *Food Research International, 29*(8), 731–750.

Iqbal, A., Khalil, I. A., Ateeq, N., & Khan, M. S. (2006). Nutritional quality of important food legumes. *Food Chemistry, 97*(2), 331–335.

Kalogeropoulos, N., Chiou, A., Ioannou, M., Karathanos, V. T., Hassapidou, M., & Andrikopoulos, N. K. (2010). Nutritional evaluation and bioactive microconstituents (phytosterols, tocopherols, polyphenols, triterpenic acids) in cooked dry legumes usually consumed in the Mediterranean countries. *Food Chemistry, 121*(3), 682–690.

Kaur, M., & Singh, N. (2005). Studies on functional, thermal and pasting properties of flours from different chickpea (Cicer arietinum L.) cultivars. *Food Chemistry, 91*(3), 403–411.

Makri, E., Papalamprou, E., & Doxastakis, G. (2005). Study of functional properties of seed storage proteins from indigenous European legume crops (lupin, pea, broad bean) in admixture with polysaccharides. *Food Hydrocolloids, 19*(3), 583–594.

Maninder, K., Sandhu, K. S., & Singh, N. (2007). Comparative study of the functional, thermal and pasting properties of flours from different field pea (*Pisum sativum L.*) and pigeon pea (*Cajanus cajan L.*) cultivars. *Food Chemistry, 104*(1), 259–267.

Meiners, C. R., Derise, N. L., Lau, H. C., Ritchey, S. J., & Murphy, E. W. (1976). Proximate composition and yield of raw and cooked mature dry legumes. *Journal of Agricultural and Food Chemistry, 24*(6), 1122–1126.

Messina, M. J. (1999). Legumes and soybeans: Overview of their nutritional profiles and health effects. *The American Journal of Clinical Nutrition, 70*(3), 439s–450s.

Miñarro, B., Albanell, E., Aguilar, N., Guamis, B., & Capellas, M. (2012). Effect of legume flours on baking characteristics of gluten-free bread. *Journal of Cereal Science, 56*(2), 476–481.

Nielsen, S. S. (1991). Digestibility of legume protein: Studies indicate that the digestibility of heated legume protein is affected by the presence of other seed components and the structure of the protein. *Food Technology, 45*(9), 112–114.

Perez-Hidalgo, M. A., Guerra-Hernández, E., & García-Villanova, B. (1997). Dietary fiber in three raw legumes and processing effect on chick peas by an enzymatic-gravimetric method. *Journal of Food Composition and Analysis, 10*(1), 66–72.

Redondo-Cuenca, A., Villanueva-Suárez, M. J., Rodríguez-Sevilla, M. D., & Mateos-Aparicio, I. (2007). Chemical composition and dietary fibre of yellow and green commercial soybeans (Glycine max). *Food Chemistry, 101*(3), 1216–1222.

Tharanathan, R. N., & Mahadevamma, S. (2003). Grain legumes—a boon to human nutrition. *Trends in Food Science & Technology, 14*(12), 507–518.

USDA Food Composition Databases. URL: https://ndb.nal.usda.gov/ndb/. Accessed on 25 July 2019.

Wani, I. A., Sogi, D. S., Wani, A. A., & Gill, B. S. (2013). Physico-chemical and functional properties of flours from Indian kidney bean (Phaseolus vulgaris L.) cultivars. *LWT-Food Science and Technology, 53*(1), 278–284.

# Chapter 2
# Introduction: Wastewater Generation

**Silu Liu and Luca Serventi**

## 2.1 Legume Consumption

Legumes (Fig. 2.1), belong to the family of *Leguminosae* and have high nutritional value. They are rich in protein, carbohydrates, fibre and minerals (Table 2.1), which may largely benefit humans' health. For example, regular consumption of legumes was confirmed to help humans avoiding various lifestyle diseases, such as diabetes and cancer (Satya et al. 2010). These health benefits are mainly attributed to the protein and minerals in legumes (Subuola et al. 2012). However, the raw seeds contain a large number of antinutritional factors (El-Haby and Habiba 2003) such as lectins, saponin, α-amylase inhibitors and protease inhibitors, hampering digestibility and bioavailability of protein and minerals (Khokhar and Apenten 2003).

Some pre-cooking processes are effective industrial methods to decrease the anti-nutrients and increase the bioavailability of minerals and protein (El-Haby and Habiba 2003). The most common industrial processes for legumes are soaking, boiling steaming, canning and sprouting (Fig. 2.2), most of these leading to solid loss (Xu and Chang 2008). The term solid loss refers to the leaching of water soluble nutrients, which include water soluble protein (Seena and Sridhar 2005), vitamins (Subuola et al. 2012), starch (Rehman and Shah 2005) and others, in the processing water. Therefore, the soaking and boiling water of legumes may contain a variety of nutrient such as protein, soluble carbohydrates and minerals. This chapter will review the main industrial methods of legume processing (soaking, boiling, steaming and canning), discussing the nutritional advantages and losses of each method.

S. Liu · L. Serventi (✉)
Department of Wine, Food and Molecular Biosciences, Faculty of Agriculture and Life Sciences, Lincoln University, Lincoln, Christchurch, New Zealand
e-mail: Luca.Serventi@lincoln.ac.nz



L. Serventi, *Upcycling Legume Water: from wastewater to food ingredients*,
https://doi.org/10.1007/978-3-030-42468-8_2

**Fig. 2.1** Seeds of five representative legume types. From left to right: haricot beans, chickpeas, green lentils, split yellow peas and yellow soybeans. (Photo: Luca Serventi, Ph.D)

**Table 2.1** The decrease in nutrients of legumes by industrial processes

| Process | Protein | Dietary fibre | Oligosaccharides | Vitamins | Phenolics | Saponins |
|---|---|---|---|---|---|---|
| Soaking | 0.7% Abd El-Hady and Habiba (2003) | 5% Rehinan et al. (2004) | 50–75% Han and Baik (2006) | 10–60% Prodanov et al. (2004) | 10–40% Xu and Chang (2008) | 8–35% Barakat et al. (2015) |
| Boiling | | 30% Rehinan et al. (2004) | 60–85% Han and Baik (2006) | 24–70% Prodanov et al. (2004) | 30–50% Xu and Chang (2008) | 40% Xu and Chang (2009) |
| Steaming | 1–5% Rehman and Shah (2005) | | | | 25% Xu and Chang (2009) | 23% Xu and Chang (2009) |
| Canning | | 16% Pedrosa et al. (2015) | 16% Pedrosa et al. (2015) | 46%–65% Słupski (2012) | | |

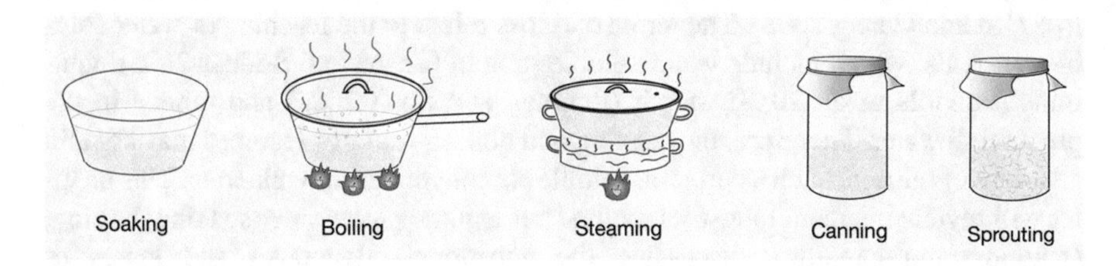


**Fig. 2.2** Diagram of soaking, boiling, steaming, canning and sprouting processes of legumes. (Credits: Dan Xiong)

## 2.2 Common Industrial Processes and Impacts

There are limited publications in the literature focused on the soaking and boiling water of legumes. However, studies on the impact of industrial process on the nutritional value of legume seeds are extensive. According to Subuola and co-authors (2012) there are two different stages in processing legumes: primary and secondary.

The primary processing stage includes sun-drying, husking, winnowing, separation and storage. The secondary processing stage refers to cleaning, soaking, blanching, boiling, steaming, roasting, fermenting, germinating, milling, sieving, frying and canning. This literature review will focus on the most common processing methods: soaking, boiling, steaming and canning. Key results are summarised in Table 2.1.

## 2.2.1 Soaking

### 2.2.1.1 Process

Traditionally, dry legume seeds need to be soaked before further processing. The raw seeds can absorb water during soaking, allowing for faster cooking and partially removing antinutritional factors. This process is frequently used to soften texture and reduce cooking time. There are differences in the soaking time of different types of legumes (Subuola et al. 2012). Generally, legumes are soaked in a 1:4 weight ratio (seed:water) between 16 and 24 hours (Han and Baik 2006). However, Subuola and co-authors (2012) argued that soaking can reduce the nutritional quality of legumes through leaching of nutrients into the soaking water.

### 2.2.1.2 Effect on Macronutrients

Han and Baik (2006) investigated the oligosaccharide content of legumes during soaking. Specifically, water soluble oligosaccharides of legumes were lost in significant amounts upon soaking. The oligosaccharides loss tended to increase with longer soaking period. The authors reported up to 50% loss in total oligosaccharides of lentils after soaking in tap water for 12 hours; this number reached 75% for chickpeas. Soaking process can lead a 56% loss in oligosaccharides of soybeans including raffinose (56% loss), stachyose (56%) and verbascose (11%).

For dietary fibres, Rehinan and collaborators (2004) reported an approximate 5% decrease in lignin of legumes after 4-hour soaking. Due to the authors employing a short soaking time period (4 hours), which is not enough for legumes to be 100% hydrated (Han and Baik 2006), therefore the dietary fibres content loss estimation of soaking process can be under estimated.

Khattab and collaborators (2009) mentioned that the soaking process can affect *in vitro* protein digestibility (IVPD) and protein content. For example, protein digestibility of cowpea and kidney beans has increased after the soaking process. This may be related to the decrease of the concentration of antinutritional factors, such as phytate, phytate phosphorus, tannins, polyphenols, trypsin and α-amylase inhibitors (Abd El-Hady and Habiba 2003). However, Abd El-Hady and Habiba (2003) stated that soaking has no significant effect on the protein digestibility of peas, chickpeas, faba and kidney beans. For example, protein digestibility of chickpeas only increased by 0.8% after soaking. Therefore, the effect of soaking on *in vitro* protein digestibility is related to the types of legumes (differences in surface area and shape of legumes). Khattab and collaborators (2009) also used two parameters (protein efficiency ratio and essential amino acid index) to show the change of protein quality. The results show that both protein efficiency ratio (PER) and essential amino acid index (EAAI) increased after soaking process and the increase in protein digestibility is influenced by the type of legumes. In addition, according to Abd El-Hady and Habiba (2003), soaking can reduce the protein content depending on the type of legumes, because some water-soluble proteins will be lost in the soaking solution. In Abd El-Hady and Habiba's experiment, the mean value of protein content of four legumes (faba beans, peas, chickpeas and kidney beans) was 26.16 g/100 g, which has decreased to 25.98 g/100 g after soaking treatment.

#### 2.2.1.3 Effect on Micronutrients

Antinutritional factors can inhibit the absorption of minerals in the human intestinal tract, such as phytic acid and polyphenols. According to Gilani and co-authors (2005), polyphenols can combine with proteins and amino acids minerals thus reducing their availability. Soaking is an effective method to prevent the binding of antinutritional factors and nutrients and reduce the content of antinutrients in order to increase the bioavailability of minerals. Luo and Xie (2014) indicated that during the soaking process of faba beans iron content decreased by 28%: from 4.7 to 3.4 mg/100 g and from 3.5 to 2.2 mg/100 g in green and white faba beans, respectively. Meanwhile, the zinc concentration of green and white faba bean decreased by 12% and 28% after the soaking process, respectively. On the other hand, the *in vitro* availability of iron and zinc in green faba beans experienced a significant increased by 18% and 6.8%; while for white faba beans, these increases were 30% and 11% for *in vitro* availability of iron and zinc in after soaking treatment, respectively. Prodanov and collaborators (2004) studied the impact of soaking and cooking on the thiamine, riboflavin and niacin contents of legumes. The authors found that soaking can significantly decrease the thiamine content of chickpeas and lentils by approximately 15% and 10%, respectively, and niacin content by 30% and 60%, respectively. Thus, authors believed that discarding the soaking liquids leads to large losses of thiamine and niacin, especially for chickpeas and lentils.

#### 2.2.1.4 Effect on Phytochemicals

Saponins are a diverse group glycosides widely distributed in the plant kingdom. Legumes are an important source of saponins. For example, soyasaponin is a kind of bioactive substance extracted from soybean that has a positive effect on human health (Augustin et al. 2011). Barakat and others (2015) studied saponin loss of faba beans and chickpeas during soaking. Authors indicated that the total amount of saponins of chickpeas and faba beans decreased by 7.7% and 35%, respectively, during 12 hours soaking. It was evident that faba beans lost more saponin than chickpeas under the same processing conditions, which may be related to the surface area and the shape of beans (Fig. 2.1). In addition, Shi and co-authors (2009) pointed that the saponins loss was related to the soaking time and the proportion of water and beans. Prolonged soaking time can increase the amount of water absorbed, soy beans can be conducive to the release of more saponins.

Furthermore, Xu and Chang (2008) studied the impact of industrial process on antioxidant activities of legumes and found decrease in legumes antioxidant activities after soaking. This is attributed to the loss in total phenolic content (TPC) of legumes after soaking. The authors observed that legumes exhibited a dramatic decrease (about 10%) in TPC after soaking in tap water for 24 hours (Xu and Chang 2008). In addition, the most significant drop in TPC was observed in lentils, with approximately a 40% loss. The different loss of total phenol content in various soybeans were dependent on the distribution of phenolic compounds in legumes. The authors confirmed that the large loss in TPC is due to the hydration and that a number of water soluble molecules of legumes can dissolve into water while soaking.

## 2.2.2 *Boiling*

### 2.2.2.1 Process

Boiling process can improve the sensory properties of legumes and reduce the heat labile anti-nutritional factors such as trypsin inhibitors (Subuola et al. 2012). According to Xu and Chang (2008), the boiling time period of pre-soaked legumes (at 100% hydration rate) is between 30 and 90 minutes and pressured cook can reduce the boiling time.

### 2.2.2.2 Effects on Macronutrients

Boiling process also has negative impact on nutrients of legumes. Han and Baik (2006) found that boiling the legumes can result in a large oligosaccharide content loss. A 30 minutes boiling of pre-soaked soybeans decreased by approximately 85% the total oligosaccharide content, while lentils exhibited an up to 60% oligosaccharide loss during the same processes.

Resistant starch has some health benefits for humans through the production of short-chain fatty acids, increased bacterial mass, and promotion of butyrate-producing bacteria (Brouns et al. 2002). Costa and co-authors (2006) studied the resistant starch content of legumes during the cooking process. Authors found that resistant starch of legumes decreased by 25% and 35% for lentils and chickpeas during cooking process (16 hours soaking and 40 minutes boiling), respectively.

Dietary fibres of legumes can also be impacted by boiling process. Rehinan and co-authors (2004) investigated the effects of soaking and cooking on dietary fibre of legumes and found large losses in fibre contents under boiling condition. A loss of fibre contents (cellulose, hemicellulose and lignin) in the extent of 30% was found for legumes after boiling. The authors concluded that even insoluble fibre of legumes can experience a large loss after boiling process.

Protein digestibility can be influenced by legume processing. The soaking process slightly increases the *in vitro* Protein Digestibility (IVPD) of legumes. In addition, the IVPD increase was even higher after the boiling process of legumes. This may due to heat-induced degradation of antinutrients. According to Han and collaborators (2007) the IVPD of legumes exhibited a dramatic increase after boiling in the distilled water at 98 °C for 30 minutes. Authors observed a 2% IVPD increase in selected legumes without pre-soaking during the boiling process; the increase in IVPD was largely depend on the type of legumes. The highest IVPD increase occurred in soybeans, accounting for approximately 12%. The authors further investigated the IVPD changes for boiled legumes with pre-soaking. Soaking the legumes in distilled water for 3 hours before boiling process can further increase the IVPD of legumes, especially for soybeans, pre-soaking can increase the IVPD of boiled soybeans by approximately 2% higher that of boiled soybeans without pre-soaking.

Similarly, Rani and others (1996) observed a significant increase in IVPD of pigeon peas after boiling in the distilled water. Boiling of the un-soaked pigeon peas can increase the IVPD by approximately 20%, while the increase in IVPD of the pre-soaked pigeon peas was over 30%. The increase in protein digestibility may due to the boiling process significantly decrease the trypsin inhibitor activity of pigeon peas. The trypsin inhibitor is an important inhibitor, which decrease the digestibility of protein, wildly exists in legumes. Authors observed that boiling the pre-soaked pigeon peas can decrease the activity of trypsin inhibitor by approximately 70%, which contributes to the increase in IVPD of boiled pigeon peas.

#### 2.2.2.3 Effect on Micronutrients

In addition, boiling can reduce vitamins contents of legumes. For example, Leskova and collaborators (2006) observed a vitamin E decrease by up to 40% during boiling of soybeans; whereas, the folate content decreased to extents of 60%. Biotin contents of legumes also were observed a small loss (5–15%) after the boiling process. Similarly, Prodonov and co-authors (2004) reported a large decrease (over 50%) in both thiamine and niacin during the boiling process.

Antinutrients in legumes can also be significantly decreased by boiling process. According to Huma and others (2008), boiling legumes can significantly decrease phytic acid and tannin contents by 55% and 35%, respectively. The phytic acid and tannin can interact with minerals; the decrease in phytic acid and tannin contents can increase the digestibility of minerals in legumes. Furthermore, Huma and others (2008) observed a mineral loss during the boiling process. The authors found up to 40% losses in minerals (including iron and zinc) for selected legumes after boiling in the distilled water. The mineral reduction was due to leaching of minerals into the boiling water.

#### 2.2.2.4 Effects on Phytochemicals

Legumes exhibited a large decrease in the total phenolic content (TPC) after boiling (Xu and Chang 2008). For example, 30 minutes boiling under 1 psi-pressure can decrease TPC of lentils by more than half. The TPC of green peas and chickpeas can reduce by 50% and 30%, respectively, after boiling for 90 minutes under 1 psi-pressure. The authors also found a relatively high TPC in the boiling water. TPC loss of legumes was significantly higher in the boiling than in the soaking condition. This may due to the impact of the temperature. On the one hand, heat increases the dissolution of TPC; approximately 35% of TPC was observed in the legume boiling water. On the other hand, part of the TPC can be destroyed by heat due to chemical reactions with other contents such as protein and starch.

Other important phytochemicals (saponins and procyanidins) can also be negatively impacted by boiling. Xu and Chang (2009) studied the impact of thermal treatments on phytochemicals in legumes including green pea, yellow pea, chickpea and lentil. Authors found that the saponin content of selected legumes can be significantly decreased during a 2-hour boiling process. For example, the loss in saponins content of chickpea exceeded 40%; approximately one third saponins contents of green pea had been decreased by boiling process. Similar to saponins, the procyanidin content of legumes also experienced a dramatic loss during the boiling process. Xu and Chang (2009) observed an average 50% loss in the procyanidin content of selected legumes after boiling in distilled water for 120 minutes. Similarly, Duhan and collaborators (2001) investigated the saponins loss during the cooking process of pigeon peas and found similar results with Xu and Chang's study. Duhan and co-authors observed a 13% loss in the saponins content of un-soaked pigeon peas after boiling process; the saponins content loss for pre-soaked pigeon peas was over 16%. The authors argued that soaking cannot significantly decrease the saponins content of pigeon peas (about 5% loss after 18 hours soaking); the decrease in saponins during the boiling process may due to the significant impact of high temperature.

### *2.2.3 Steaming*

#### 2.2.3.1 Process

Steaming is an alternative method for cooking legumes. Steaming uses less amount of water than boiling, helping the nutrients retention of legumes during cooking process. According to Xu and Chang (2009), common methods of legume steaming require boiling water for 70 minutes. In the industrial practise, autoclaving (a pressure steaming method) is used to decrease the steaming time period to approximately 20 minutes (Rehman and Shah 2005) due to higher pressure generates higher temperature (over 120 °C) of water vapour.

#### 2.2.3.2 Effect on Macronutrients

In Rehman and Shah (2005)'s study, the authors selected different types of legumes to investigate the effects of thermal heat treatment on macronutrients, including protein content, protein digestibility, starch content and starch digestibility. Thermal heat processing lead to the reduction of protein content in different legumes (black gram, chickpeas, lentils, red kidney beans and white kidney beans). Protein content of these legumes decreased by 1.3–4.6% upon cooking in autoclave at 121 °C for 10–90 minutes and at 128 °C for 20 minutes. In addition, the decrease of protein content was directly proportional to the autoclaving temperature and time. For example, the protein content in black grams decreased by 0.5% (from 24.7% to 24.2%) upon autoclaving at 121 °C for 10 minutes, with loss increasing to 0.9% (from 24.7% to 23.8%) when the treatment was extended to 90 minutes.

Rehman and Shah (2005) also investigated that the impact of thermal heat treatment on the protein digestibility of selected legumes including black grams, red kidney beans, white kidney beans, chickpeas and lentils, determining that protein digestibility can be significantly increased (doubled) after the autoclaving process. The change of protein digestibility depends on two factors: steam temperature and time period of autoclaving. At the same temperature (121 °C), prolonging the autoclaving time period (from 10 minutes up to 90 minutes) decreased the protein digestibility. Meanwhile, the digestibility of protein was inversely proportional to temperature at the same time.

Lastly, both starch content and starch digestibility can be affected by thermal heat method. Rehman and Shah (2005) mentioned that starch content of selected legumes decreased by 1.1–6.7% during steaming process in the temperature of 121–128 °C. Consistently with changes in protein content, starch decreased over steaming time. On the other hand, Rehman and Shah (2005) also indicated that autoclaving treatment is an important factor affecting the starch digestibility. Starch digestibility of raw legumes was 36–42% and increased to 87–91% after thermal heat method. Steaming may have increased the hydrolysis rate of starch and decreased the content of anti-nutritional factors.

#### 2.2.3.3 Effects on Phytochemicals

Phytochemicals, such as phenolics, procyanidins, saponins, phytic acids and phenolic acids, can largely benefit for human health. However, steaming process can significantly decrease phytochemicals in legumes. For example, steaming at 100 °C for 70 minutes can decrease the total phenolic content of selected legumes by 25% (Xu and Chang 2009) contributing to a decrease in antioxidant capacity. Large losses in procyanidin and phytic acid contents were also found. Selected legumes, after a 70-minute steaming treatment, exhibited an over 20% average loss in procyanidin contents and an approximately 17% loss in phytic acid contents (Xu and Chang 2009). Similarly, Rehman and Shah (2005) found that thermal heat can lead to phytic acid decrease. Phytic acid is one of the antinutritional factors, which can cause reduced uptake in minerals and protein. The decrease in phytic acid contents of legumes is relatively large at approximately 50% after a 90 minutes autoclaving at 121 °C. The large decrease in phytic acid significantly increases the digestibility of proteins and minerals such as iron and zinc. Thus, steaming may help to avoid protein and starch losses when processing legumes.

Saponins are among the important phytochemicals, expressing several bioactivities. The steaming treatment had a negative impact on saponin contents of legumes. According to Xu and Chang (2009), up to 23% of saponin contents can be lost during the steaming process. The significant loss in saponin contents may partially due to the heat instability of saponins and the leaching into steaming water. Ozcan and co-authors (2014) analysed the benefits of phenolic acids and announced that phenolic acids are one of the important antioxidant agents and protective agents of DNA against free radicals. Legumes contain phenolic acids that benefit human health; however, the streaming process may decrease the phenolic acid content of food legumes (Xu and Chang 2009). Up to 38% in phenolic acid content (for chickpeas) can be lost during the steaming treatment and the average loss in phenolic acid contents of legumes is approximately 10%. The oxidative degradation of phenolic acid such as enzymatic browning is one of the important reasons that the phenolic acid content decreased during streaming (Xu and Chang 2009). The formation of phenolic acid–protein complexes and phenolic acid–tannin complexes during streaming may also contribute to the decrease of phenolic acid content (Ozcan et al. 2014).

## *2.2.4 Canning*

### 2.2.4.1 Process

Canning is a technological process that allows cooking of legumes packaged in cans. The canning technology is used for food preservation and provides legumes in all seasons. According to Uebersax (2006) canned beans have the characteristics of bright colour, convenient edible and long storage period. However, this process will significantly increase legumes' cost. In addition, human may concern about the texture, colour or nutritional value of beans will change in the process of canning (Parmar et al. 2016). Canning consists of a series of processes. Before the final packaging, legumes should be soaked for 18 hours at room temperature, then heated in a water bath at 80 °C and blanched for 4 minutes. Later, beans and brine (1.3% weight/volume salt and 1.6% weight/volume sugar) are sealed in cans and treated at 141 °C for 14 minutes (Nleya et al. 2011). Parmar and collaborators (2016) described that the canning procedure includes soaking (25 °C for 12 hours), blanching (85 °C for 30 minutes), canning in brine (1.3% salt and 1.6% sugar) and final heating (121 °C for 14 minutes).

#### 2.2.4.2 Effect on Macronutrients

Pedrosa and co-authors (2015) investigated the effects of industrial canning on the nutritional composition of Spanish dry beans (Almonga and Curruquilla bean). Canning significantly increased the content of protein, ash, starch and dietary fibre, while fat and carbohydrates decreased after the canning process. For example, protein content of Almonga and Curruquilla increased by 6.8% and 24% after canning, respectively. The reason may be the loss of soluble nutrients. In addition, according to Costa and others (2006) the reason for the increase of dietary fibre content is related to the formation of fibre compounds, such as protein-fibre complexes. Parmar and collaborators (2016) investigated the effect of canning on soluble sugar (sucrose), α-galactoside (stachyose) and ciceritol in Almonga and Curruquilla. The concentration of sucrose decreased by 57% (Almonga) and 59% (Curruquilla) after the canning process. In addition, stachyose and ciceritol experienced a significant decrease by 37% (stachyose) and 64% (ciceritol) in canned Almonga; while the reduction of stachyose and ciceritol of canned Curruquilla was 50% and 82%, respectively. The reduction of sucrose, stachyose and ciceritol was attributed to leaching form legumes into the canning water.

#### 2.2.4.3 Effect on Phytochemicals

Canning not only can affect physical characteristics of legumes, such as weight, volume, bulk density, hydration capacity, hydration index, swelling capacity, swelling index and colour, but also decrease the concentration of various phenolic compounds. For example, the average concentration of catechin, chlorogenic acid, protocatechuic acid, vanillic acid, ferulic acid and sinapic acid in different legumes decreased after canning, by 0.21, 0.27, 1.36, 0.09, 0.55 and 0.22 mg/g, respectively (Parmar et al. 2016). On contrast, the content of gallic acid increased by 0.09 mg/g during canning. The reason why phenolic compounds (except gallic acid) decreased after canning treatment may be related to leaching into the brine. Conversely, the increase in gallic acid increase may be due to monomer depolymerization or compound release from tannins.

Canning can also reduce the concentration of trace elements. Pedrosa and co-authors (2015) determined that the content of trace elements such as calcium, phosphorous and magnesium significantly decreased during industrial canning process. These trace elements can be lost in the canning liquid. Results showed that iron loss (70%) was 2.3 times higher than that of zinc (30%), possibly related to the distribution of iron and zinc in the seeds and the molecular structure of the beans (Afify et al. 2011).

Słupski (2012) studied the impact of canning on the content of thiamine and riboflavin and the change after stored for 12 month in five types canned beans (Mona, Alamo, Flaforte, Igolomska and Laponia). Both the thiamine and riboflavin content decreased after canning for 12 months. The content of thiamine decreased by up to 63% in selected beans; while the content of riboflavin decreased by 50%,

18%, 50% and 60% in Mono, Alamo, Flaforte and Igolomska, respectively. Only the canned Laponia exhibited a slight increase (4.3%) after 12 months' storage (Słupski 2012). The bean type and storage period can significantly affect the content of thiamine and riboflavin. On the one hand, the change of the content is related to the type of bean due to the degree of seed coat cracking that can lead to more loss. On the other hand, the longer the storage time, the more soluble substances are lost in canning water. However, the increase of riboflavin may be attributed to the prevention of vitamin degradation by starch and protein (Nisha et al. 2005).

## 2.3 Conclusions

Legumes are rich in nutrients and are abundantly consumed all around the world. Nonetheless, legumes contain some anti-nutritional factors that have adverse effects on the digestion, absorption and utilization of nutrients and produce adverse physiological reactions. Cooking methods (soaking, boiling, steaming and canning) not only can be used to reduce adverse effects of anti-nutritional factors, but also have a positive effect on improving the absorption and utilization of nutrients. However, wastewater is generated and nutrients are leached in it at relevant amounts.

**Acknowledgments** The completion of this book chapter was made possible thanks to the funding allocated to the taught Master course "FOOD 698 – Research Essay" by Lincoln University.

## References

Abd El-Hady, E. A., & Habiba, R. A. (2003). Effect of soaking and extrusion conditions on antinutrients and protein digestibility of legume seeds. *LWT – Food Science and Technology, 36*, 285–293.

Afify, A. E. M. M., El-Beltagi, H. S., El-Salam, S. M. A., & Omran, A. A. (2011). Bioavailability of iron, zinc, phytate and phytase activity during soaking and germination of white sorghum varieties. *Plos one, 6*(10).

Augustin, M., Kuzina, V., Andersen, B., & Bak, S. (2011). Molecular activities, biosynthesis and evolution of triterpenoid saponins. *Phytochemistry, 72*(6), 435–457.

Barakat, H., Reim, V., & Rohn, S. (2015). Stability of saponins from chickpea, soy and faba beans in vegetarian, broccoli-based bars subjected to different cooking techniques. *Food Research International, 76*, 142–149.

Brouns, F., Kettitz, B., & Arrigoni, E. (2002). Resistant starch and "the butyrate revolution". *Trends in Food Science & Technology, 13*(8), 251–261.

Costa, G., Queiroz-Monici, K., Reis, S., & Oliveira, A. (2006). Chemical composition, dietary fibre and resistant starch contents of raw and cooked pea, common pea, chickpea and lentil legumes. *Food Chemistry, 94*(3), 327–330.

Duhan, A., Khetarpaul, N., & Bishnoi, S. (2001). Saponin content and trypsin inhibitor activity in processed and cooked pigeon pea cultivars. *International Journal of Food Sciences and Nutrition, 52*, 53–59.

El-Haby, E. A., & Habiba, R. A. (2003). Effect of soaking and extrusion conditions on antinutrients and protein digestibility of legume seeds. *Swiss Society of Food Science and Technology, 36*, 285–293.

Gilani, S., Cockell, A., & Sepehr, E. (2005). Effects of antinutritional factors on protein digestibility and amino acid availability in foods. *Journal of AOAC International, 88*, 967–987.

Han, I. H., & Baik, B. K. (2006). Oligosaccharide content and composition of legumes and their reduction by soaking, cooking, ultrasound, and high hydrostatic pressure. *Cereal Chemistry, 83*(4), 428–433.

Han, I. H., Swanson, B. G., & Baik, B. K. (2007). Protein digestibility of selected legumes treated with ultrasound and high hydrostatic pressure during soaking. *Cereal Chemistry, 84*(5), 518–521.

Huma, N., Anjum, F. M., Sehar, S., Khan, M. I., & Hussain, S. (2008). Effect of soaking and cooking nutritional quality and safety of legumes. *Nutrition and Food Science, 38*(6), 570–577.

Khattab, R. Y., & Arntfeld, S. D. (2009). Nutritional quality of legume seeds as affected by some physical treatments 2. Antinutritional factors. *LWT – Food Science and Technology, 42*, 1113–1118.

Khokhar, S., & Apenten, R. K. O. (2003). Antinutritional factors in food legumes and effects of processing. *The Role of Food, Agriculture, Forestry and Fisheries in Human Nutrition, 4*, 82–116.

Leskova, E., Kubikova, J., Kovacikova, E., Kosicka, M., Porubska, J., & Holcikova, K. (2006). Vitamin losses: Retention during heat treatment and continual changes expressed by mathematical models. *Journal of Food Composition and Analysis, 19*, 252–276.

Luo, Y., & Xie, W. (2014). Effect of soaking and sprouting on iron and zinc availability in green and white faba bean (Vicia faba L.). *Journal of Food Science and Technology, 51*, 3970–3976.

Nisha, P., Singhal, R. S., & Pandit, A. B. (2005). A study on degradation kinetics of riboflavin in green gram whole (Vigna radiata L.). *Food chemistry, 89*(4), 577–582.

Nleya, T., Arganosa, G., Vandenberg, A., & Tyler, R. (2011). Genotype and environment effect on canning quality of kabuli chickpea. *Canadian Journal of Plant Science, 82*, 267–272.

Ozcan, T., Akpinar-Bayizit, A., Yilmaz-Ersan, L., & Delikanli, B. (2014). Phenolics in human health. *International Journal of Chemical Engineering and Applications, 5*(5), 393–396.

Parmar, N., Singh, N., Kaur, A., Virdi, A., & Thakur, S. (2016). Effect of canning on color, protein and phenolic profile of grains from kidney bean, field pea and chickpea. *Food Research International, 89*(1), 526–532.

Pedrosa, M., Cuadrado, C., Burbano, C., Muzquiz, M., Cabellos, B., Olmedilla-Alonso, B., & Asensio-Vegas, C. (2015). Effects of industrial canning on the proximate composition, bioactive compounds contents and nutritional profile of two Spanish common dry beans (*Phaseolus vulgaris L.*). *Food Chemistry, 166*(1), 68–75.

Prodanov, M., Sierra, I., & Vidal-Valverde, C. (2004). Influence of soaking and cooking on the thiamin, riboflavin and niacin contents of legumes. *Food Chemistry, 84*, 271–277.

Rani, S., Jood, S., & Sehgal, S. (1996). Cultivar differences and effect of pigeon pea seeds boiling on trypsin inhibitor activity and in vitro digestibility of protein and starch. *Nahrung, 40*(3), 145–146.

Rehinan, Z., Rashid, M., & Shah, W. H. (2004). Insoluble dietary fibre components of food legumes as affected by soaking and cooking processes. *Food Chemistry, 85*, 245–249.

Rehman, Z., & Shah, W. H. (2005). Thermal heat processing effects on antinutrients, protein and starch digestibility of food legumes. *Food Chemistry, 91*, 327–331.

Satya, S., Kaushik, G., & Naik, S. N. (2010). Processing of food legumes: A boon to human nutrition. *Mediterranean Journal of Nutrition and Metabolism, 3*(3), 183–195.

Seena, S., & Sridhar, K. R. (2005). Physicochemical, functional and cooking properties of under explored legumes, *Canavalia* of the southwest coast of India. *Food Research International, 38*, 803–814.

Shi, J., Xue, J., Ma, Y., Li, D., Kakuda, Y., & Lan, Y. (2009). Kinetic study of saponins B stability in navy beans under different processing conditions. *Journal of Food Engineering, 93*(1), 59–65.

Słupski, J. (2012). Effect of freezing and canning on the thiamine and riboflavin content in immature seeds of five cultivars of common bean (Phaseolus vulgaris L.). *International Journal of Refrigeration 35*(4):890–896.

Subuola, F., Widodo, Y., & Kehinde, T. (2012). Processing and utilization of legumes in the tropics. In *Trends in vital food and control engineering* (pp. 71–84). InTech: Rijeka, Croatia.

Uebersax, M. (2006). Dry edible beans: Indigenous staple and healthy cuisine. *Forum on Public Policy: A Journal of the Oxford Round Table,* pp. 1-27.

Xu, B., & Chang, S. K. C. (2008). Effect of soaking, boiling and steaming on total phenolic content and antioxidant activities of cool season food legumes. *Food Chemistry, 110*(1), 1–13.

Xu, B., & Chang, S. K. C. (2009). Phytochemical profiles and health-promoting effects of cool-season food legumes as influenced by thermal processing. *Journal of Agriculture and Food Chemistry, 57*, 10718–10731.

# Chapter 3
# Soaking Water Composition

**Luca Serventi**

## 3.1 Soaking of Legumes

Legumes have become important parts of the human and animal diet, providing the protein, fat, dietary fibre and minerals essential to the body, also as a substitute for expensive animal protein. They belong to the family of *Leguminosae* and can be listed into three sub families: Caesalpinieae, Mimosoideae, and Papilionoideae (Rehinan et al. 2004). Legumes mainly include beans, chickpeas, lentils, peas and soy. Due to their active role in the treatment and prevention of diabetes, cardiovascular disease and lowering blood cholesterol, legumes are widely planted worldwide (Prodanov et al. 2004).

Although legumes have health promoting properties, they often require soaking or other forms of processing to soften the seed before cooking and remove antinutrients (Naviglio et al. 2013). Different types of legumes require different soaking times, as well any additional ingredient in the soaking water. Eating unprocessed legumes often leads to indigestion, mainly because they contain fiber-related anti-nutritional factors such as polyphenols, α-galactosides, trypsin and chymotrypsin inhibitors, phytate and lectin. These factors can adversely affect protein digestion, mineral absorption and starch digestion (Shi et al. 2018). Therefore, removing these factors from legumes can improve their nutritional value. Existing techniques such as soaking and cooking can effectively reduce the anti-nutritional factors in the legume (Costa et al. 2018).

In previous studies, it was found that soaking and cooking of beans with water reduced their nutrient content, with phytic acid and tannin being the most significant, while the highest reduction was observed after soaking with sodium

L. Serventi (✉)
Department of Wine, Food and Molecular Biosciences, Faculty of Agriculture and Life Sciences, Lincoln University, Lincoln, Christchurch, New Zealand
e-mail: Luca.Serventi@lincoln.ac.nz



L. Serventi, *Upcycling Legume Water: from wastewater to food ingredients*,
https://doi.org/10.1007/978-3-030-42468-8_3

bicarbonate. These treatments also resulted in varying degrees of reduction in nutrients such as protein, minerals and total sugar. Lestienne and collaborators (2005) observed a significant reduction in phytic acid content (17–28%) by soaking millet, corn, rice and soybeans at 30 °C for 24 hours. At the same time, they also observed that iron was reduced by 60% in soaked rice, and zinc was reduced by 30% in millet and corn. Moreover, El-Hady and Habiba's experiment (2003) found that the average loss of protein content in the soaking process of four beans (broad bean, pea, chickpea and cowpea) was 0.18 g/100 g (from 26.16 to 25.98 g/100 g). It indicates that the water-soluble protein in the beans is transferred to the solution during the soaking process. From these findings it can be inferred that the liquids used to treat legumes contain large amounts of nutrients, which has great potential for producing more economical and high-quality food ingredients.

This chapter summarizes the composition of legume soaking water: macronutrients, mineral profile and antinutrients. Minerals, phytic acid and trypsin inhibitor were quantified in legume soaking water in the experimental sections of this chapter: 3.4.1 and 3.4.2.

## 3.2 Macronutrients

Huang and collaborators (2018) soaked 5 different legumes (haricot beans, chickpeas, whole green lentils, split yellow peas and yellow soybeans) at a ratio of 1:3.3 (dried beans:water) for 16 hours at room temperature. Soaked legumes were then drained and their wastewater collected for further analysis. The experimental results showed that haricot beans and split yellow peas released the highest amount of solids in the bean soaking water: 2.38 and 1.89 and g/100 g, respectively (Fig. 3.1). These results suggested that a high amount of solids was easily leached from beans and peas, possibly due to their location on the seed coat (Sreerama et al. 2010), to the split geometry of peas or the high amount of soluble carbohydrates present.

The highest amounts of water soluble carbohydrates were found in beans and peas: 0.65 and 0.69 g/100 g (Huang et al. 2018), representing about one third of the dry matter (Fig. 3.1). Several cultivars of beans (Winged, Great Northern, California small white, Red, Navy, Pinto, Pink, Black eye, Bengal, Mung, Red and Broad), peas (smooth, wrinkled, cowpea) were found to contain significantly more soluble carbohydrates (soluble fibre and sugars) than lentils and soy. Specifically, the sugar content of navy beans and wrinkled peas was found to be 5.6–6.2 and 10–15 g/100 g, respectively, higher than 4.2–6.1 g/100 g of lentils and 5.3 g/100 g of soy (Reddy et al. 1984). The highest amount of soluble fibre was found in the soaking water of peas, in agreement with studies on legumes showing that peas contain more soluble and less insoluble fibre than chickpeas and lentils (McCleary 2008; Tosh and Yada 2010). Studies have shown that soluble carbohydrates of legumes are distributed in the inner fraction of the husk, mainly in the cotyledons; consisting of cell wall polysaccharides with different levels of water solubility, such as oligosaccharides

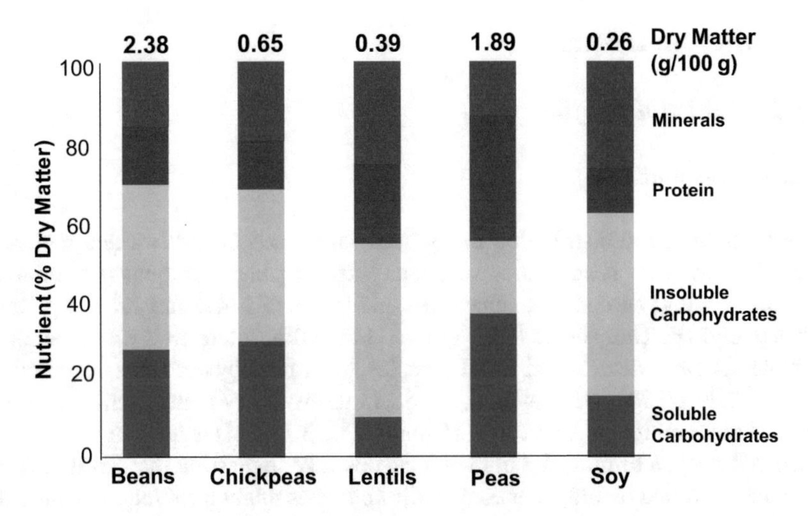


**Fig. 3.1** Nutrient distribution in legume soaking water (% dry matter), data from Huang and collaborators (2018)

(alpha-galactosidase, raffinose, stachyose and verbascose) and pectin (McCleary 2008; Tosh and Yada 2010).

Insoluble carbohydrates represent the major fraction, consisting of about 40% of the dry matter, with the exception of peas (about 20%) (Fig. 3.1). In terms of quantity, beans released more insoluble carbohydrates into the soaking water. Beans are known to contain more insoluble fibre than other legumes: 20–28 vs. 10–18 g/100 g (Tosh and Yada 2010). These insoluble carbohydrates are mainly found in the outer layer of the husk. Therefore, fragmentations and physical damage of the seeds may increase this specific loss. Interestingly, the relative fraction of insoluble carbohydrates per dry matter was high in beans, chickpeas and peas (Fig. 3.1). While for beans it could be due to their high content, results of chickpeas might be explained by their irregular shape, while for peas it could be due to the split structure which compromised the integrity of the outer husk. The main constituent of insoluble carbohydrates found in legumes are three polysaccharides: cellulose, hemicelluloses and pectin (Tosh and Yada 2010). It is therefore hypothesized that at least one of these three fibres leached in the soaking water of legumes. Further studies might warrant the exact fibre composition as this might explain the physicochemical properties of these liquids.

Proteins were found in relevant amounts (0.35–0.60 g/100 g) in the soaking water of beans and peas (Huang et al. 2018). Proportionally, protein represented a high fraction only for pea water: about 30% of it (Fig. 3.1). Protein is mainly located in the endosperm, which is found in the inner core of legume seeds (Kawamura 1967). Split peas present an exposed endosperm. Therefore, it is reasonable to deduce that the fractionation in half of peas resulted in higher protein leaching upon soaking.

## 3.3 Micronutrients

### *3.3.1 Mineral Profile*

#### 3.3.1.1 Introduction

Mineral content and distribution in legumes vary based on cultivar and region. In general, common beans and soybeans contain high amount of minerals (4.2–5.1 g/100 g) compared to chickpeas and lentils (2.9–4.0 and 2.4–4.1 g.100 g, respectively) (El Tinay et al. 1989; Lin and Lai 2006). Interestingly, high amounts of minerals were determined in all samples, with the highest value recorded for beans at 0.39 g/100 g (Huang et al. 2018). In terms of dry matter, lentils and peas released more minerals: about 30% of solids (Fig. 3.1). This result can be attributed to the distribution of minerals in the seeds, possibly suggesting that lentil minerals were mainly found on the outer coat, while split peas might have released more due to their exposed endosperm. A detailed mineral profile of legume soaking water was not available. Therefore, this experimental section (3.4.1) outlines the key minerals quantified in these ingredients.

#### 3.3.1.2 Materials and Methods

Legume soaking water was freeze-dried and then allowed to analysis of their mineral profile. Legumes tested included: haricot beans (Sun Valley Foods, New Zealand), chickpeas (Kelley Bean Co, NE, USA), green lentils (McKenzie's, Australia), split yellow peas (Cates, New Zealand) and yellow soybeans (YZ NON-GMO BEAN, Sunson, New Zealand). Samples were prepared with a microwave digestor (CEM MARS Xpress (CEM Corporation, Matthews, NC, USA). After preparation, absorbance was measured spectrophotometric technique (Agilent, Mulgrave, Victoria, Australia) by ICP-OES (Inductively Coupled Plasma Optical Emission Spectrophotometer). Results were expressed as means and standard deviations of three replicates. Statistical analysis was performed via one-way ANOVA (Minitab), p value 0.05.

#### 3.3.1.3 Results and Discussion

Minerals are important nutritionally as well as for food quality, they are an important aspect to be considered when taking sensory analysis due to their effects on taste (Whelton et al. 2007). The mineral profile of legume soaking water is shown in Table 3.1, displaying the content per dry matter. The ash fraction was dominated by potassium for all samples evaluated, ranging from 6407 mg/100 g (lentils) to 9249 (beans) mg/100 g (Table 3.1), representing about 75–80% of all minerals. The second group of minerals consisted of phosphorous, sulphur, magnesium, calcium and, particularly in chickpeas, sodium, with values in the range of 100–1000 mg/100 g

**Table 3.1** Mineral content of legume soaking water (mg/100 g dry matter)

| Mineral content of legume soaking water (mg/100 g dry matter) | Beans | Chickpeas | Lentils | Peas | Soy |
|---|---|---|---|---|---|
| Potassium (K) | 9249[b] | 9901[a] | 6407[d] | 6361[d] | 7800[c] |
| Phosphorous (P) | 729[d] | 721[d] | 816[c] | 923[b] | 1093[a] |
| Sulphur (S) | 476[b] | 308[d] | 492[b] | 641[a] | 342[c] |
| Magnesium (Mg) | 453[b] | 307[c] | 266[d] | 226[e] | 694[a] |
| Calcium (Ca) | 268[c] | 295[b] | 123[e] | 244[d] | 480[a] |
| Sodium (Na) | 81.8[b] | 170[a] | 46.0[c] | 80.0[b] | 34.5[d] |
| Zinc (Zn) | 6.70[b] | 3.84[d] | 3.24[e] | 13.7[a] | 5.94[c] |
| Iron (Fe) | 5.11[b] | 5.15[b] | 1.21[c] | 4.89[b] | 18.4[a] |
| Copper (Cu) | 1.59[b] | 1.61[b] | 0.94[d] | 2.25[a] | 1.21[c] |
| Molybdenum (Mo) | 1.25[b] | 0.62[c] | 0.04[e] | 0.13[d] | 1.36[a] |
| Manganese (Mn) | 0.57[d] | 11.4[a] | 0.90[cd] | 2.00[b] | 1.52[bc] |

Different letters refer to statistically significant difference ($p < 0.05$)

(Table 3.1). The third group was represented by zinc, iron, copper, molybdenum and manganese, with values from 0.04 mg/100 g (molybdenum in lentils) to 13.7 mg/100 g (zinc in peas) (Table 3.1). The nutritional relevance of these nutrients is based on their recommended daily intake (RDI). Therefore, discussion of these profiles is based on nutritional recommendations.

When considering 100 g of dry matter from the soaking water, all legumes tested provided amounts of potassium, phosphorous, sulphur, magnesium and calcium that exceeded the RDI (Table 3.1). Potassium levels were as high as 2 times the RDI (9249 and 9901 mg/100 g, respectively) for beans and chickpeas (Table 3.1) (Ministry of Health New Zealand 2019). Significant differences were determined across legumes. Chickpeas were the best source for potassium and manganese (261% and 88% RDI, respectively) (Ministry of Health New Zealand 2019), covering about 2 times the RDI. Sodium levels were the highest in chickpeas, possibly affecting sensory quality if used as food ingredient, but only delivering 8.5% of the maximum sodium intake recommended (Ministry of Health New Zealand 2019). Peas were the best source of sulphur, zinc and copper, providing to 43%, 114% and 132% RDI, respectively (Ministry of Health New Zealand 2019). Sulphur is associated to several amino acids in proteins so this result was likely due to the higher protein content of this soaking water (Fig. 3.1). Nonetheless, soy represented the richest source of a variety of minerals, including phosphorous, magnesium, calcium, iron and molybdenum (Table 3.1). The results of iron and molybdenum were significant. The dry matter from soaking of yellow soybeans contained the equivalent of 102% and 230% the RDI for women and men, respectively (Ministry of Health New Zealand 2019). Considering that iron deficiency is a serious and growing nutritional issue (Camaschella 2019) these results are particularly relevant to the nutritional community. In addition, results for molybdenum were astonishing. The molybdenum content ranged from 0.04 g/100 mg of lentils to 1.36 mg/100 g of soy (Table 3.1), which equal to 131% and 3989% of its RDI, so from 1 to 40 times the amount recommended (Ministry of Health New Zealand 2019). Molybdenum is an important cofactor in the enzymatic catabolism of aminoacids (Turnlund et al. 1995) and its occurrence is highly affected by the soil type (James et al. 2019).

### *3.3.2 Phytochemicals and Vitamins*

A low amount of phytochemicals was found in the soaking water of beans, lentils and peas: about 0.3 mg/100 g of phenolics and 3 mg/100 g of saponins (Huang et al. 2018). Previous studies (Segev et al. 2011; Xu and Chang 2008) revealed a phenolic loss upon soaking, which supports these findings. Both saponins and phenolics are known to be bound to protein in legumes (Güçlü-Üstündağ and Mazza 2007), therefore are mainly found in the seed endosperm. Higher phytochemical loss from beans and peas can be explained with their higher amount of protein leach (0.35 and 0.60 g/100 g vs. 0.0–0.08 g/100 g) (Huang et al. 2018), while in the case of whole green lentils it could be due to the higher phenolic content inherent to lentils, about 5 times higher than for other legumes (Dueñas et al. 2016).

## 3.4 Antinutrients

### *3.4.1 Phytic Acid*

#### 3.4.1.1 Introduction

Phytic acid also is a storage form of phosphorous in legume seeds. Its structure chelates minerals such as calcium, iron and zinc, thus limiting their bioavailability (McKie and Mc Cleary 2016). Phytate content in legumes ranges from 0.55% to 1.55% (Lott et al. 2000). Soaking has been shown to reduce phytate content in legumes by up to 55%, being particularly effective at high temperature (45–65 °C) and mildly acidic pH (Csapó and Albert 2018; Dueñas et al. 2016) (Gupta et al. 2015). Therefore, it was hypothesised that legume soaking water may contain phytic acid.

#### 3.4.1.2 Materials and Methods

Phytic acid content of legume wastewater was quantified with an enzymatic assay. Alkaline phosphatase suspension (1.2 mL), buffer (25 mL, pH 5.5), sodium azide (0.02% w/v), phytase suspension (1.2 mL), buffer (25 mL, pH 10.4), magnesium chloride, zinc sulfate and sodium azide (0.02% w/v), phosphorus standard solution (24 mL, 50 μg/mL) and sodium azide (0.02% w/v) were supplied as a part of the total phosphorus and phytic acid kit obtained from Megazyme. Ammonium molybdate, ascorbic acid, distilled water and ascorbic acid were used as well. A colour reagent was prepared in 3 steps:

- A first solution consisted of 10 g ascorbic acid dissolved in 90 mL distilled water by stirring, followed by the addition of 5.35 mL of concentrated sulphuric acid, to a final volume of 100 mL reached with distilled water;
- A second solution consisted of 1.25 g ammonium molybdate dissolved in 20 mL distilled water and final volume of 25 mL obtained with distilled water;
- Finally, the colour reagent was obtained by combining one part of the second solution with five parts of the first solution.

The extraction of phytic acid from legume soaking was based on a "total phosphorus" and "free phosphorus" principle. For "free phosphorous", 0.5 mL of sample were combined with 0.62 mL of distilled water and 0.20 mL of phytase assay buffer. For "total phosphorus", only 0.05 mL of sample were combined with 0.60 mL of distilled water and 0.20 mL of phytase assay buffer. Solutions were incubated at 40 °C for 10 minutes, then diluted with 0.02 mL of distilled water and 0.20 mL of ALP (alkaline phosphatase) assay buffer. ALP assay buffer (0.20 mL) and alkaline phosphatase suspension (0.02 mL) were added to the total phosphorus reaction. Both solutions were mixed and incubated at 40 °C for 15 minutes, then diluted with 0.30 mL of trichloroacetic acid (50% w/v) and centrifuged at 13,000 rpm for 10 minutes. The supernatant (1 mL) was transferred to colorimetric determination (Greiner et al. 2002). Briefly, 0.5 mL of colour reagent were added to 1 mL of supernatant, vortexed and incubated in a water bath set at 40 °C for 1 hour. Later, absorbance was recorded at 655 nm. The absorbance values of samples and phosphorus standard solutions were used in the calculation of total phosphorus and phytic acid. Phosphorous calibration curve was prepared with concentrations ranging from 0 to 7.5 μg/ΔAphosphorus. Calculations of total phosphorus and phytic acid followed the method described in the literature by McKie and Mc Cleary (2016). The phytic acid content was calculated as follows and expressed in g/100 g:

$$Total\ phosphorous = \left( \left( Mean\ M * M * V \right) / \left( \frac{F}{10000} * w * v \right) \right) * \Delta \text{Aphosphorus}$$

$$Phytic\ acid = \frac{Total\ phosphorous}{0.282}$$

Where:

Mean M = mean value of phosphorus standards (μg/ΔAphosphorus)
V = sample volume (mL)
F = dilution factor
ΔA = absorbance change of sample
10,000 = conversion from μg/g to g/100 g
w = sample weight (g)
v = sample volume

#### 3.4.1.3 Results and Discussion

Legumes are soaked to allow softening and faster cooking as well as to remove antinutrients. Phytic acid in particular is known to reduce the absorption of minerals such as calcium, iron and zinc (Schlemmer et al. 2009). Typical levels of phytic acid (expressed in g/100 g dry matter) in legumes were found to be relatively high, in comparison to other foods: 0.51–1.77 for navy beans, 0.28–1.60 chickpeas, 0.22–1.22 peas, 0.27–1.51 lentils and 0.92–1.67 for cooked soybeans (Kumar et al. 2010; Schlemmer et al. 2009). Therefore, one may expect high levels of phytate. On the contrary, this study proved that the amount of phytic acid leached into soaking water was negligible. Lentils and pea soaking water were the highest source of phytic acid, contained about 0.04 g/100 g (Fig. 3.2). These results showed that approximately 3–15% of the phytate typically present in lentils and peas was lost in the soaking water. On the contrary, virtually no leaching was observed for the other legumes. For example, soy soaking water contained 0.021 g/100 g of phytic acid (Fig. 3.2), equal to a 1–2% loss. Recent studies (Shi et al. 2018; Wang et al. 2010) showed that soaking alone does not significantly affect phytate content of legumes, but rather soaking in alkaline solutions or cooking does and, even so, to a low extent (Huma et al. 2008; Shi et al. 2018). The difference among legumes could be explained with their geometry. Phytic acid is found in the inner cotyledon of legume seeds, therefore it is hardly in contact with the soaking water. Lentils and peas were, to a small extent, exceptions, due to their thin structure (lentils) and split geometry (peas) that allowed higher interaction between inner cotyledon and processing water.

### 3.4.2 *Trypsin inhibitors*

#### 3.4.2.1 Introduction

Legumes contain proteins known as trypsin inhibitors, which are involved in protecting the plant protection against animals. They are mainly located in the seed coat and in the embryo (Avilés-Gaxiola et al. 2018). Two main families were

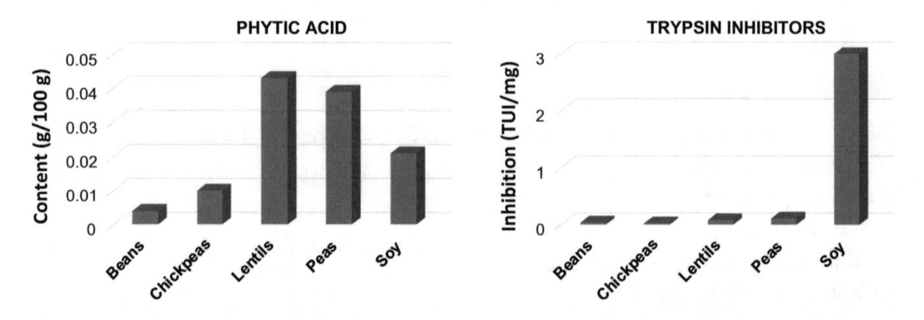


**Fig. 3.2** Contents of phytic acid (left) and trypsin inhibitors (right) of legume soaking water

classified: Kunitz and Bowman-Birk (BBI) (Guillamon et al. 2008). The Bowman-Birk inhibitors of chickpeas, lentil, peas and soybeans were determined to inhibit carcinogenesis in the gastrointestinal tract by acting as protease inhibitors, thus protecting serine from carcinogenic mechanisms (Clemente and del Carmen Arques 2014). Nonetheless, trypsin inhibitors exert negative activities as well, primarily by inhibiting the absorption of sulphur-rich proteins such as trypsin and chymotrypsin, resulting in limited growth (Csapó and Albert 2018; Pesic et al. 2007). It has been reported that food processing, mainly thermal treatments, could inactivate trypsin inhibitors. Therefore, quantification of these antinutrients was performed on legume soaking water.

#### 3.4.2.2 Materials and Methods

Trypsin inhibitors were determined with a colorimetric method (Napoleão et al. 2013). A 1 mL aliquot was taken from each sample and destabilized by adding an equal volume of assay buffer with vigorous shaking for 2 minutes, followed by filtration with Whatman No.2 filter paper. This step allowed binding of trypsin to a chromogenic substrate called BAPA (benzoyl-*DL*-arginine-*p*-nitroanilide hydrochloride), which produced coloured product. In the presence of inhibitors, trypsin binds them, resulting in no colour. Samples were diluted to different volumes in order to obtain 30–70%. A 2 mL aliquot of BAPA substrate (40 mg of BAPA in 10 mL of dimethyl sulfoxide) was mixed with 0.5 mL of trypsin solution, then incubated at 37 °C for 10 minutes. The reaction was stopped upon addition of 0.5 mL of 30% acetic acid. Absorbance was read at 410 nm. The content of trypsin inhibitors was calculated as trypsin units inhibited (TUI) per milligram of sample with the following equation:

$$Trypsin\,Inhibitors = \left(A410standard - A410sample\right) * 100 / \left(V * C\right)$$

Where:

A = Absorbance (AU)
V = sample volume (mL)
C = sample concentration (mg/mL)

#### 3.4.2.3 Results and Discussion

Raw legumes contain variable amounts of trypsin inhibitors (expressed as TUI/mg legume seed): 16 for navy beans, 8.1–16 chickpeas, 5.1–7.4 green lentils, 2.2–4.2 split yellow peas, and 46–94 soybeans (Avilés-Gaxiola et al. 2018; Shi et al. 2017; Valdebouze et al. 1980). These antinutrients inhibit the activity of trypsin, which is a protease involved in protein digestion. Therefore, these chemicals are unwanted in foods. The experiment performed revealed that only soybeans leached trypsin inhibitors into the soaking water, to a level of 3 TUI/mg (Fig. 3.2). The other legumes released little to none: 0–0.1 TUI/mg (Fig. 3.2). Previous studies showed that soaking reduced the content of trypsin inhibitors to minor extents: 6% for red beans (Patterson et al. 2017), 9% navy beans, lentils, chickpeas, and 19% soybeans (Shi et al. 2017). The only drastic reduction was registered for split yellow peas (30%) (Shi et al. 2017) possibly due to its split geometry that allowed exposure of the proteinaceous endosperm to the processing water, therefore increasing the release of these antinutritive factors. The results of the experiment disagree with values from the literature for peas and soy. It is possible that different cultivars carry a diverse range and amount of inhibitors. Nonetheless, both studies on soaked legumes (Patterson et al. 2017; Shi et al. 2017) and the study on legume soaking water (Fig. 3.2) highlighted minor leaching of these antinutrients into the soaking water. Therefore, these ingredients carry insignificant amounts of trypsin inhibitors (0.0–3.0 TUI/mg) in comparison to other foods such as tofu and soymilk (1.1–7.0 and 1.6–14 TUI/mg, respectively) (Murugkar 2014). In case no trypsin inhibitor is wanted in food, heat treatment at 75 °C for 1 hour has been proven successful (Friedman and Gumbmann 1986).

## 3.5 Conclusions

The first step in legume processing is soaking. A study on legume soaking water revealed nutritionally important fractions of soluble carbohydrates, up to 0.69 g/100 g for split yellow peas (Huang et al. 2018) equivalent to 37% of the dry matter (Fig. 3.1). Lower levels of insoluble carbohydrates and proteins were found. Interestingly, high amounts of minerals leached into the processing water, up to 0.39 g/100 g for haricot beans (Huang et al. 2018), 16% of the dry matter (Fig. 3.1). The most abundant mineral was potassium (9901 mg/100 g dry matter for chickpeas) (Table 3.1), but high levels were recorded also for others. A 100 g dose of solids from legume soaking water could contribute to 1–2 times the recommended daily intake for potassium, phosphorous, magnesium and iron. On the contrary, the amount of phytochemicals (phenolics and saponins) and antinutrients (phytic acid and trypsin inhibitor) was low. If these liquids were to be dried into a powder, the amount of antinutrients may become significant. For example, soy soaking water contained 3 TUI/mg of trypsin inhibitor, with only 0.26 g/100 g of dry matter. Therefore, a solid powder would contain about 1000 TUI/mg. These results may suggest the use of legume soaking water for the extraction of soluble fibre and minerals. Minerals could be directed to the industry of nutritional supplements. Soluble fibre could promote physicochemical properties such as foaming, emulsifying and thickening. In addition, oligosaccharides from legumes might exert prebiotic activity, favouring to the growth of health-promoting lactobacilli (Yang et al. 2011). These properties have been investigated and results are discussed in the next chapter: Chap. 4.

**Acknowledgments** The author acknowledges Roger Cresswell and Lynne Clucas for analysing the mineral profile of legume soaking water. The author also thank Kaviya Sathyanarayanan for quantifying phytic acid and Lirisha Vinola Dsouza for analysing trypsin inhibitor and Letitia Stipkovits for planning their experimental design. Funding was provided by Lincoln University in support of the courses "FOOD 699 – Research Placement". Finally, acknowledgments go to Mingyu Chen and Xiong Dan for contributing to the scientific discussion of carbohydrates and minerals, respectively.

## References

Avilés-Gaxiola, S., Chuck-Hernández, C., & Serna Saldivar, S. O. (2018). Inactivation methods of trypsin inhibitor in legumes: A review. *Journal of Food Science, 83*(1), 17–29.

Camaschella, C. (2019). Iron deficiency. *Blood, 133*(1), 30–39.

Clemente, A., & del Carmen Arques, M. (2014). Bowman-Birk inhibitors from legumes as colorectal chemopreventive agents. *World journal of gastroenterology: WJG, 20*(30), 10305.

Costa, R., Fusco, F., & Gândara, J. F. (2018). Mass transfer dynamics in soaking of chickpea. *Journal of Food Engineering, 227*, 42–50.

Csapó, J., & Albert, C. (2018). Methods and procedures for reducing soy trypsin inhibitor activity by means of heat treatment combined with chemical methods. *Acta Universitatis Sapientiae, Alimentaria, 11*(1), 58–80.

Dueñas, M., Sarmento, T., Aguilera, Y., Benitez, V., Mollá, E., Esteban, R. M., & Martín-Cabrejas, M. A. (2016). Impact of cooking and germination on phenolic composition and dietary fibre fractions in dark beans (*Phaseolus vulgaris L.*) and lentils (*Lens culinaris L.*). *LWT-Food Science and Technology, 1*(66), 72–78.

El Tinay, A. H., Mahgoub, S. O., Mohamed, B. E., & Hamad, M. A. (1989). Proximate composition and mineral and phytate contents of legumes grown in Sudan. *Journal of Food Composition and Analysis, 2*(1), 69–78.

El-Hady, E. A., & Habiba, R. (2003). Effect of soaking and extrusion conditions on antinutrients and protein digestibility of legume seeds. *LWT-Food Science and Technology, 36*(3), 285–293.

Friedman, M., & Gumbmann, M. R. (1986). Nutritional improvement of soy flour through inactivation of trypsin inhibitors by sodium sulfite. *Journal of Food Science, 51*(5), 1239–1241.

Greiner, R., Larsson Alminger, M., Carlsson, N. G., Muzquiz, M., Burbano, C., Cuadrado, C., Pedrosa, M. M., & Goyoaga, C. (2002). Pathway of dephosphorylation of myo-inositol hexakisphosphate by phytases of legume seeds. *Journal of Agricultural and Food Chemistry, 50*(23), 6865–6870.

Güçlü-Üstündağ, Ö., & Mazza, G. (2007). Saponins: properties, applications and processing. *Critical reviews in food science and nutrition, 47*(3), 231–258.

Guillamon, E., Pedrosa, M. M., Burbano, C., Cuadrado, C., de Cortes Sánchez, M., & Muzquiz, M. (2008). The trypsin inhibitors present in seed of different grain legume species and cultivar. *Food Chemistry, 107*(1), 68–74.

Gupta, R. K., Gangoliya, S. S., & Singh, N. K. (2015). Reduction of phytic acid and enhancement of bioavailable micronutrients in food grains. *Journal of Food Science and Technology, 52*(2), 676–684.

Huang, S., Liu, Y., Zhang, W., Dale, K. J., Liu, S., Zhu, J., & Serventi, L. (2018). Composition of legume soaking water and emulsifying properties in gluten-free bread. *Food Science and Technology International, 24*(3), 232–241.

Huma, N., Anjum, M., Sehar, S., Issa Khan, M., & Hussain, S. (2008). Effect of soaking and cooking on nutritional quality and safety of legumes. *Nutrition & Food Science, 38*(6), 570–577.

James, B., Ting, J., & Wang, G. (2019). Molybdenum (Mo) availability in soil, dietary intake and its health risk assessment in the soil-food crops system. *Environment international.*

Kawamura, S. (1967). Quantitative paper chromatography of sugars of the cotyledon, hull, and hypocotyl of soybeans of selected varieties. *Technical Bulletin of Faculty of Agriculture, 18* (2), 117–131.

Kumar, V., Sinha, A. K., Makkar, H. P., & Becker, K. (2010). Dietary roles of phytate and phytase in human nutrition: A review. *Food Chemistry, 120*(4), 945–959.

Lestienne, I., Icard-Vernière, C., Mouquet, C., Picq, C., & Trèche, S. (2005). Effects of soaking whole cereal and legume seeds on iron, zinc and phytate contents. *Food Chemistry, 89*(3), 421–425.

Lin, P. Y., & Lai, H. M. (2006). Bioactive compounds in legumes and their germinated products. *Journal of Agricultural and Food Chemistry, 54*(11), 3807–3814.

Lott, J. N., Ockenden, I., Raboy, V., & Batten, G. D. (2000). Phytic acid and phosphorus in crop seeds and fruits: A global estimate. *Seed Science Research, 10*(1), 11–33.

McCleary, N. (2008). Technological aspects of dietary fibre. In *Advanced dietary fibre technology*. Oxford: Blackwell Science.

McKie, V. A., & Mc Cleary, B. V. (2016). A novel and rapid colorimetric method for measuring total phosphorus and phytic acid in foods and animal feeds. *Journal of AOAC International, 99*(3), 738–743.

Ministry of Health New Zealand. Nutrient reference values. URL: https://www.nrv.gov.au/nutrients. Accessed on 01 Aug 2019.

Murugkar, D. A. (2014). Effect of sprouting of soybean on the chemical composition and quality of soymilk and tofu. *Journal of Food Science and Technology, 51*(5), 915–921.

Napoleão, T. H., dos Santos-Filho, T. G., Pontual, E. V., da Silva Ferreira, R., Coelho, L. C. B. B., & Paiva, P. M. G. (2013). Affinity matrices of Cratylia mollis seed lectins for isolation of glycoproteins from complex protein mixtures. *Applied Biochemistry and Biotechnology, 171*(3), 744–755.

Naviglio, D., Formato, A., Pucillo, G. P., & Gallo, M. (2013). A cyclically pressurised soaking process for the hydration and aromatisation of cannellini beans. *Journal of Food Engineering, 116*(3), 765–774.

Patterson, C. A., Curran, J., & Der, T. (2017). Effect of processing on antinutrient compounds in pulses. *Cereal Chemistry, 94*(1), 2–10.

Pesic, M., Vucelic-Radovic, B., Barac, M., Stanojevic, S., & Nedovic, V. (2007). Influence of different genotypes on trypsin inhibitor levels and activity in soybeans. *Sensors, 7*(1), 67–74.

Prodanov, M., Sierra, I., & Vidal-Valverde, C. (2004). Influence of soaking and cooking on the thiamin, riboflavin and niacin contents of legumes. *Food Chemistry, 84*(2), 271–277.

Reddy, N. R., Pierson, M. D., Sathe, S. K., & Salunkhe, D. K. (1984). Chemical, nutritional and physiological aspects of dry bean carbohydrates—a review. *Food Chemistry, 13*(1), 25–68.

Rehinan, Z.-U., Rashid, M., & Shah, W. (2004). Insoluble dietary fibre components of food legumes as affected by soaking and cooking processes. *Food Chemistry, 85*(2), 245–249.

Schlemmer, U., Frølich, W., Prieto, R. M., & Grases, F. (2009). Phytate in foods and significance for humans: Food sources, intake, processing, bioavailability, protective role and analysis. *Molecular Nutrition & Food Research, 53*(S2), S330–S375.

Segev, A., Badani, H., Galili, L., Hovav, R., Kapulnik, Y., Shomer, I., & Galili, S. (2011). Total phenolic content and antioxidant activity of Chickpea (Cicer arietinum L.) as affected by soaking and cooking conditions. *Food and Nutrition Sciences, 2*(7), 724.

Shi, L., Mu, K., Arntfield, S. D., & Nickerson, M. T. (2017). Changes in levels of enzyme inhibitors during soaking and cooking for pulses available in Canada. *Journal of Food Science and Technology, 54*(4), 1014–1022.

Shi, L., Arntfield, S. D., & Nickerson, M. (2018). Changes in levels of phytic acid, lectins and oxalates during soaking and cooking of Canadian pulses. *Food Research International, 107*, 660–668.

Sreerama, Y. N., Neelam, D. A., Sashikala, V. B., & Pratape, V. M. (2010). Distribution of nutrients and antinutrients in milled fractions of chickpea and horse gram: Seed coat phenolics and their distinct modes of enzyme inhibition. *Journal of Agricultural and Food Chemistry, 58*(7), 4322–4330.

Tosh, S. M., & Yada, S. (2010). Dietary fibres in pulse seeds and fractions: Characterization, functional attributes, and applications. *Food Research International, 43*(2), 450–460.

Turnlund, J. R., Keyes, W. R., Peiffer, G. L., & Chiang, G. (1995). Molybdenum absorption, excretion, and retention studied with stable isotopes in young men during depletion and repletion. *The American Journal of Clinical Nutrition, 61*(5), 1102–1109.

Valdebouze, P., Bergeron, E., Gaborit, T., & Delort-Laval, J. (1980). Content and distribution of trypsin inhibitors and hemagglutinins in some legume seeds. *Canadian Journal of Plant Science, 60*(2), 695–701.

Wang, N., Hatcher, D. W., Tyler, R. T., Toews, R., & Gawalko, E. J. (2010). Effect of cooking on the composition of beans (*Phaseolus vulgaris L.*) and chickpeas (*Cicer arietinum L.*). *Food Research International, 43*(2), 589–594.

Whelton, A. J., Dietrich, A. M., Burlingame, G. A., Schechs, M., & Duncan, S. E. (2007). Minerals in drinking water: Impacts on taste and importance to consumer health. *Water Science and Technology, 55*(5), 283–291.

Xu, B., & Chang, S. K. (2008). Effect of soaking, boiling, and steaming on total phenolic content and antioxidant activities of cool season food legumes. *Food Chemistry, 110*(1), 1–13.

Yang, B., Prasad, K. N., Xie, H., Lin, S., & Jiang, Y. (2011). Structural characteristics of oligosaccharides from soy sauce lees and their potential prebiotic effect on lactic acid bacteria. *Food Chemistry, 126*(2), 590–594.

# Chapter 4
# Soaking Water Functional Properties

**Luca Serventi, Congyi Gao, Wendian Chang, Yaying Luo, Mingyu Chen, and Venkata Chelikani**

## 4.1 Introduction

To the best of our knowledge, only one paper has previously investigated the functional properties of legume soaking water (Huang et al. 2018). This study determined the emulsifying properties of the soaking water from five legumes. Results showed interesting functionality for haricot beans, chickpeas and split yellow peas, with an emulsifying ability of 40–50%, while lower values were observed for green lentils (about 20%) and yellow soybeans (6%). The emulsifying properties of legume soaking water were attributed to multiple factors: protein, ratio of soluble to insoluble carbohydrates, and saponins (Huang et al. 2018). Legume protein, soluble carbohydrates and saponins are known to exert multiple functionalities: foaming, emulsifying and thickening (Jarpa-Parra 2018; Lafarga et al. 2019; Liu et al. 2017; Rehal et al. 2019; Shevkani et al. 2019; Wang et al. 2019).

Furthermore, oligosaccharides of legumes have been proven effective as prebiotics, by supporting the growth of probiotic lactobacilli (Johnson et al. 2015; McSwain et al. 2019; Mohanty et al. 2018; Siva et al. 2019). In addition, bioactives found in legume soaking water, such as phenolics, saponins and peptides (Cilla et al. 2018; Lopez-Martinez et al. 2017; Mojica and de Mejia 2018), exert antimicrobial activities which could prevent the growth of pathogenic bacteria such as *E. coli* (Kanatt et al. 2011; Sitohy and Osman 2010).

With increasing consumers interests in gastrointestinal and digestive health, prebiotic and probiotic foods are being developed owing to the association of benefits to hosts (Caballero et al. 2015). Prebiotics include non-digestible carbohydrates and proteins, for examples, resistant starch, undigested oligosaccharides, and non-starch polysaccharides (Swennen et al. 2006). Legume seeds contain a significant amount

L. Serventi (✉) · C. Gao · W. Chang · Y. Luo · M. Chen · V. Chelikani
Department of Wine, Food and Molecular Biosciences, Faculty of Agriculture and Life Sciences, Lincoln University, Lincoln, Christchurch, New Zealand
e-mail: Luca.Serventi@lincoln.ac.nz



L. Serventi, *Upcycling Legume Water: from wastewater to food ingredients*,
https://doi.org/10.1007/978-3-030-42468-8_4

**Fig. 4.1** Appearance of legume soaking water. From left to right: haricot beans, chickpeas, green lentils, split yellow peas and yellow soybeans. (Photo by Luca Serventi, Ph.D.)

of raffinose family oligosaccharides, which are potential prebiotics based on their α(1–6) links (Moussou et al. 2017). Probiotics are living microorganisms which confer benefits on the human body (Gibson and Roberfroid 1995). The most common probiotics are genera *Lactobacillus*, *Enterococcus*, *Saccharomyce*s and *Bifidobacterium* (Caballero et al. 2015). The benefits of probiotics include: reduced blood cholesterol, inhibiting the growth of pathogens, enhancing the immune system, and relieving allergy and lactose intolerance, among others (Caballero et al. 2015). However, these microorganisms are hard to sustain and replicate in the colon and gastrointestinal tract due to the low environmental pH and competition with existed microbes (Swennen et al. 2006). Prebiotics are stable in human digestible tract and alternatively stimulate the growth of probiotics in the colon (Martinez-Villaluenga et al. 2008). Thus, prebiotics can mediate the immune system by altering intestinal microbiota (Schley and Field 2002).

Therefore, it is possible that the by-products of legume soaking (Fig. 4.1) could exert one or more of these functionalities. Consequently, a new study has been performed to obtain a comprehensive characterisation of the physicochemical properties of legume soaking water: protein solubility, pH, foaming ability, emulsifying activity, water absorption and oil absorption. In addition, the microbiological potential of these by-products has been evaluated, focusing on two key effects: prebiotic and antimicrobial activities.

## 4.2 Materials and Methods

### *4.2.1 Protein Solubility*

Five legumes were tested: haricot beans (Sun Valley Foods, Auckland, New Zealand), chickpeas (Kelley Bean Co., Scottsbluff, NE, USA), green lentils (McKenzie's, Victoria, Australia), split yellow peas (Cates, Ashburton, New

Zealand) and yellow soybeans (Sunson Asian Food Market, Wigram, Christchurch, New Zealand). Soaking was performed with the method described by Huang and collaborators (2018). Briefly, 3.3 parts of water were added to 1 part of dry legume, then soaking was conducted over a 16-hour period at room temperature. Later, water was drained from the legumes and analysed. The degree of solubility of proteins found in legume soaking water was evaluated with the Bradford protein essay (Kruger 2009). A 10 ml aliquot of soaking water was pipetted into 15 ml falcon tubes and centrifuged at 12000 rpm for 30 minutes at 4 °C. After centrifugation, the supernatant was collected and transferred to another falcon tube. Bovine Serum Albumin (Sigma-Aldrich, St. Louis, Mo, USA) was used as a standard. Dye reagent was prepared (1:4 ratio of protein assay dye reagent concentrate (Bio-Rad, Ca, USA) and water) and filtered using Whatman no. 1 filter paper. Each standard or sample (50 μl) was mixed with 2.5 ml of dye reagent in 3 ml cuvettes and left to incubate at room temperature for at least 5 minutes. The absorbance was measured at 595 nm using a V-1200 spectrophotometer (VWR, Radnor, PA, USA).

### 4.2.2 *pH*

The pH of legume soaking water was measured with a pH meter (Mettler Toledo, SevenEasy pH, Switzerland).

### 4.2.3 *Foaming Ability*

Each soaking water was freeze-dried prior to the test of foaming ability. Solutions of 100 ml were prepared by dissolving 1 g of freeze-dried powder into 100 ml of water, as by published method (Sathe and Salunkhe 1981). Whipping of solutions was performed with a benchtop mixer (500A Delta Food Equipment, Canada) set at speed 5 for 7 minutes. Foaming ability was determined with the following equation:

$$\mathrm{FA}\ (\%) = \frac{V_f - V_i}{V_i} * 100$$

Where

FA = stands for Foaming Ability (expressed as percentage)
$V_f$ = stands for final Volume
$V_i$ = stands for initial Volume

### 4.2.4 Emulsifying Activity Index

Freeze-dried powders obtained from legume soaking water were tested for the emulsifying activity index based on the method developed by Pearce and Kinsella (1978). In synthesis, 0.4 g of powder were dissolved in 20 ml of water and 20 ml of soybean oil (Pams, New Zealand) into 50 ml centrifuge tubes. Later, samples were homogenized for 1 minute at speed 4 (DuPont Instruments, Model Omni Mixer 17,106). Then, 20 μl aliquots of emulsions were diluted in 5 ml of SDS (Fischer, New Zealand), prior to vortex mix and measurement of absorbance at 500 nm (V-1200 spectrophotometer, VWR, Radnor, Pennsylvania, USA). Results were calculated based on the formula by Jiang and collaborators (2009):

$$\mathrm{EAI} = \left( \frac{4.606 * \mathrm{Absorbance}}{\mathrm{DM} * \mathrm{VF} * 1000} \right) * \mathrm{DF}$$

Where

EAI = stands for Emulsifying Activity Index
4.606 is a constant factor
Absorbance was measured in AU
DM = is the dry matter of the water solution (0.02 g/g)
1000 is a constant factor
VF = stands for volume fraction of oil (0.5)
DF = stands for dilution factor (250)

### 4.2.5 Water- and Oil-Absorption Capacity

Dried powders were tested for thickening abilities with a modified version of a published method (Kaur and Singh 2005). Fractions of 1 g of each powder were dissolved in 8 ml of either water or soybean oil (Pams, New Zealand) by hand shaking and vortex shaking them in 15 ml centrifuge tubes. Subsequently, centrifugation was achieved (Heraeus Multifuge XIR, Thermo Scientific, Massachusetts, USA) with 10 minutes at speed of 1260 g. Once centrifuged, the supernatant was discarded and the weight of the remaining pellet and tube was measured. Water Absorption Capacity (WAC)) and Oil Absorption Capacity (OAC)) were calculated as the ration of pellet weight by sample weight, expressed as g/g.

### 4.2.6 Prebiotic Activity

Prebiotic properties of the legume soaking water were determined by adding 900 μl of soaking water samples to 100 μl of 0.8 OD *Lactobacillus acidophilus* cell suspension into 1.5 ml Eppendorf tubes. These samples were incubated at 37 °C for

24 hours anaerobically and after incubation, samples were serially diluted by a factor of $10^{-15}$ with 0.1% peptone water. 100 µl of these diluted samples were spread on the *Lactobacilli* MRS agar plates. The experiments were carried out in duplicates and CFU (colony forming units) were determined.

### 4.2.7 Antimicrobial Activity

The antimicrobial activity of soaking water was tested on indicator organism using selective media MacConkey agar plates. Plates were spread with 100 µl 0.8 OD *E. coli* cell suspension and then the agar surface was dug out of the same size circular hole having diameter of 1.5 cm. Samples of each legume soaking water (1.5 ml) were poured into the wells, while an antibiotic disc was placed in the spare area for comparison. Experiments were carried out in duplicates and inhibition zones were measured.

### 4.2.8 Statistical Analysis

All analyses were performed in triplicate, with the exception of the microbiological tests (Sects. 4.3.6 and 4.2.7) that were performed in duplicate. Statistical analysis was performed with Minitab®17, by one-way ANOVA and Tukey's post hoc test (p value 0.05).

## 4.3 Results and Discussion

### 4.3.1 Protein Solubility and pH

All samples tested primarily consisted of soluble protein: ranging from 66% to 100% of the protein fraction in peas and lentils, respectively (Fig. 4.2). Pulses and soybeans contain protein of various solubility based on their secondary structure and media (Singhal et al. 2016). Protein loss in soaking water was merely due to leaching, with no external factors such as heating, mechanical actions or use of solvents. Therefore, it is plausible that most of the nutrient found in it was highly soluble in water. The only exception was found for split yellow peas: protein solubility of about 66% (Fig. 4.2). As their name implies, yellow peas were split in half, thus exposing the endosperm to the soaking water. This exposure likely caused the leaching of protein with low water solubility. In addition, the presence of hydrophilic units is crucial to water solubility. Values of media pH far from the protein isoelectric point (around 4.0–5.0) resulted in the highest solubility, reaching 60–80%

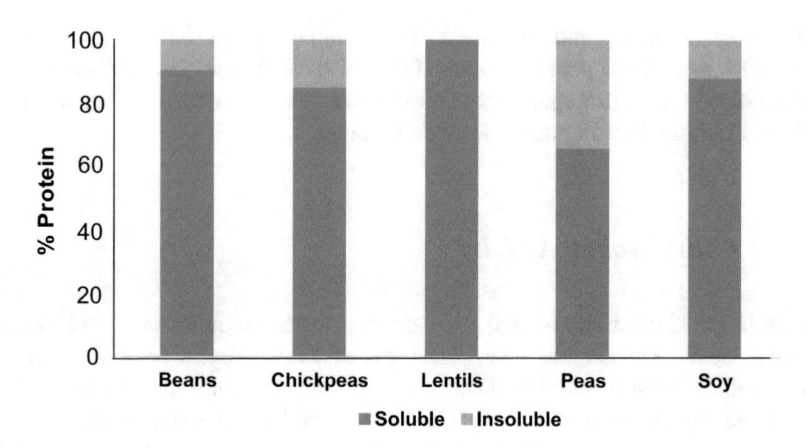


**Fig. 4.2** Protein solubility of legume soaking water

at pH 7.0 for cowpeas, kidney beans and peas (Shevkani et al. 2019). The soaking water tested were characterized by the following pH values: 6.7 (beans), 6.5 (chickpeas), 5.7 (lentils), 6.1 (peas) and 6.3 (soy). Therefore, most of the ingredients tested were near neutral in terms of pH, which favoured protein solubility. Unlike other legumes, lentil soaking water exhibited high protein solubility despite a pH value close to isoelectric point (5.7). Previous research demonstrated superior solubility for protein isolates from lentils in comparison to those from chickpeas and soy (Jarpa-Parra 2018) likely explaining these findings.

### *4.3.2 Foaming Ability*

Table 4.1 shows the foaming and emulsifying activity of freeze dried powders from legume soaking water. Peas and soybeans exerted the strongest foaming ability: 19% vs. 4.0–7.3% (Table 4.1). These results were drastically lower than those reported for the cooking water of the same legumes: 38–97% (Serventi et al. 2018; Stantiall et al. 2018). It must be observed that this test analysed 1% solutions, while data literature on cooking water was based on solutions ranging from 3% to 5% (Serventi et al. 2018; Stantiall et al. 2018). Even after accounting for dry matter, the results of the soaking water are lower than the cooking counterpart. The foaming ability of legume is mainly related to the protein content of the ingredients. According to Zayas (1997), foams are defined as two-phase systems of air cells separated by thin continuous liquid layers called the lamellar phases. When air is injected in the soaking water foam bubbles are produced. This system is not stable, thus surfactants are needed between at the air and water interface. Surfactants migrate toward such interface during whipping, lowering the free energy of soaking water, then creating a two-phase system (Makri et al. 2005).

**Table 4.1** Foaming and emulsifying abilities of freeze-dried legume soaking water

| Sample | Foaming ability (%) | Emulsifying activity index ($m^2/g$) |
|---|---|---|
| Haricot beans | 7.3 ± 2.3[b] | 5.3 ± 1.2[c] |
| Chickpeas | 6.7 ± 3.1[b] | 15 ± 0[b] |
| Green lentils | 4.0 ± 3.5[b] | 20 ± 1[b] |
| Split yellow peas | 19 ± 5[a] | 50 ± 7[a] |
| Yellow soybeans | 19 ± 1[a] | 18 ± 1[b] |

Both tests were conducted on 1% solutions of water (foaming) and water-oil (emulsifying)
Different letters indicate statistically different treatments

When analysing the solid content of the soaking water, it appears that peas contained more protein than the other pulses (Huang et al. 2018) thus explaining their higher foaming ability. Nonetheless, soy contained very low mounts of protein and yet foamed to the same extent as peas (Table 4.1). Soy performance could be ascribed to different amino acidic profile and protein conformation between soy and pulses. The main peptides found in soybeans were glycinin (11S) and β-conglycinin (7S) (Mills et al. 2001). On the contrary, the main peptides found in pulses like peas were albumin (2S), vicilin (7S) and legumin (11S) (O'Kane et al. 2005). Different peptide fractions and ratios result in significant changes to the foaming ability of legume protein (Klupšaitė and Juodeikienė 2015). According to the Ribeiro and collaborators (2013), foaming ability is also affected by temperature, sodium concentration and pH of the aqueous phase (Mitra and Dungan 2000), as well as the presence of saponins (Mitra and Dungan 1997). The pH of the solutions tested did not vary significantly (Sect. 4.3.1). The sodium content was extremely high for chickpeas (170 mg/100 g dry matter) and low for soybeans (35 mg/100 g dry matter) while the other samples contained 46–82 mg/100 g dry matter (Chap. 3). Sodium can change protein conformation, resulting in altered foam stability (Raikos et al. 2007). These differences might further support the excellent performance of soybeans (Table 4.1), despite low protein content (Huang et al. 2018).

### *4.3.3 Emulsifying Ability*

Split yellow peas expressed the highest ability of emulsification: 50 $m^2/g$ versus an average of 15–20 $m^2/g$ of the other legumes. The only exception was represented by haricot beans, with minor activity: 5.3 $m^2/g$ (Table 4.1). Studies on legume cooking water showed a comparable range of EAI values: 15–47 $m^2/g$ (Damian et al. 2018; Serventi et al. 2018). What differed from cooking water was the ranking of the ingredients. While the best emulsifier among cooking water was lentils, the best one among soaking water was peas.

According to Dickinson (2003), emulsification and foaming abilities are related to polysaccharides for their water-holding and oil-holding properties, with a role played by proteins as well. In the case of proteins, as being amphiphilic, adsorption proceeds through sequential attachment of several polypeptide segments (Mcwatters and Cherry 1977). However, unlike foaming ability, the emulsification effect of soybeans was not as satisfactory. In the researches by Sosulski and McCurdy (1987), soybeans expressed higher EAI than peas. Soy proteins may interact with lipids, reducing their hydrophilicity and stability emulsion conformational stability (Makri et al. 2005; Tolstoguzov 1997). Furthermore, salinity participates to the emulsifying ability. Specifically, NaCl addition changes protein structure, possibly destabilizing the emulsion (Raikos et al. 2007). Finally, phytochemical with surfactant abilities (esterified phenolics and most saponins) can stabilise emulsions (Bordenave et al. 2014; Güçlü-Üstündağ and Mazza 2007). Moderate levels of saponins were found in the soaking water of beans, lentils and peas (Huang et al. 2018) thus not fully explaining the results observed. High ratio of soluble carbohydrates to dry matter seemed to be the likely rationale behind the best emulsifying activities observed.

### *4.3.4 Thickening Abilities*

Water and oil absorption capacities of ingredients influence food quality and stability. Hence, WAC and OAC were tested. All powders showed preference for oil: OAC values ranging from 2.10 to 2.73 g/g, whereas WAC data spanned from 0.22 to 0.71 g/g (Table 4.2). Haricot powder showed the lowest WAC and OAC: 0.22 g/g and 2.10 g/g, respectively. At the same time, chickpea powder expressed the highest WAC (0.71 g/g) while pea powder expressed the highest OAC (2.73 g/g).

It was surprising to observe almost no water absorption, given the promising results by the cooking water of the same legumes. Damian and co-authors (2018) reported a WAC of 1.5 g/g for chickpea cooking water and 2.2 for that of split yellow peas. Similarly, Serventi and co-authors (2018) determined a WAC of 1.5 g/g for the cooking water of yellow soybeans. The soaking water tested were characterised by very low density, from 1.01 to 1.10 g/ml. It is possible that the soluble solids found in soaking water were not fully separated during centrifugation, thus being lost with the supernatant removal. Further experiments at higher centrifugation speeds must be conducted to verify this hypothesis. Legume soaking water contained both soluble and insoluble carbohydrates (Huang et al. 2018). These polymers influence WAC and OAC of their powdered ingredients. Sila and collaborator

**Table 4.2** Water and oil absorption capacities of freeze-dried legume soaking water

| Sample | Water absorption capacity (g/g) | Oil absorption capacity (g/g) |
|---|---|---|
| Haricot beans | $0.22 \pm 0.07^{e}$ | $2.10 \pm 0.10^{e}$ |
| Chickpeas | $0.71 \pm 0.02^{a}$ | $2.43 \pm 0.12^{b}$ |
| Green lentils | $0.45 \pm 0.06^{c}$ | $2.33 \pm 0.06^{c}$ |
| Split yellow peas | $0.47 \pm 0.09^{b}$ | $2.73 \pm 0.15^{a}$ |
| Yellow soybeans | $0.40 \pm 0.03^{d}$ | $2.17 \pm 0.20^{d}$ |

Different letters indicate statistically different treatments

(2014) reported that oligosaccharides and other water-soluble polysaccharides promote higher hydrophilic interactions.

Generally speaking, legume flours have higher affinity for oil than water, due to the abundance of insoluble macronutrients (Sumnu et al. 2016). Therefore, higher OAC than WAC were expected. Oil affinity of soaking water was comparable to that of cooking water. For example, the OAC of chickpeas was 2.4 g/g for soaking water (Table 4.2) and 3.2 g/g for cooking water (Damian et al. 2018). Legume proteins and insoluble fibre can absorb oil. Because cooking water contained more protein and insoluble fibre than soaking (yet to low extent), lower OAC was expected. Protein in particular can promote hydrophobic interaction, particularly when of low solubility. As shown in Fig. 4.1, peas contained a higher fraction of insoluble protein than other samples: 66% vs. 85–100% (Fig. 4.1). Consequently, the soaking water of peas and other legumes can act affectively as thickener in fat-based systems.

### 4.3.5 *Prebiotic Activity*

Testing the effect of legume soaking water on *Lactobacillus* growth revealed fascinating results. Green lentils halted the growth of this probiotic bacteria, resulting in no CFU at all dilutions tested, comparably to peptone water (Table 4.3). On the contrary, the other legume soaking water significantly enhanced probiotic growth, with values of $6.4 \times 10^{-16}$ CFU (split yellow peas) and $8.0 \times 10^{-16}$ CFU (chickpeas) (Table 4.3). Remarkably, haricot beans and yellow soybeans were so effective that colony count was not possible: TMC (Table 4.3).

Legume derived oligosaccharides have been shown to promote probiotic growth (Moussou et al. 2017; Swennen et al. 2006). A previous study have shown significant amounts of these water soluble carbohydrates in the soaking water of certain legumes, ranging from 0.19 g/100 g of chickpeas to 0.65 and 0.69 g/100 g of haricot beans and split yellow peas (Huang et al. 2018). These nutritional profile might explain the extremely positive result of haricot beans and the lower number for

**Table 4.3** Number of CFU (Colony Forming Units) of *lactobacillus* grew in plates incubated with legume soaking water

| | *Lactobacillus* (CFU) | | |
|---|---|---|---|
| Dilutions | $10^{-6}$ | $10^{-12}$ | $10^{-15}$ |
| Peptone water | 0 | 0 | 0 |
| Nutrient broth | TMC | 17 | 7 |
| Haricot beans | TMC | TMC | TMC |
| Chickpeas | TMC | TMC | 80 |
| Green lentils | 0 | 0 | 0 |
| Split yellow peas | TMC | 101 | 64 |
| Yellow soybeans | TMC | TMC | TMC |

Dilutions of $10^{-6}$, $10^{-12}$ and $10^{-15}$. The term TMC means too many colonies to count chickpeas but not the lower value for split yellow peas and the high proliferation for yellow soybeans (Table 4.3). Lentils was the only sample tested to not promote any growth. It has been shown that a peptide found in lentils, called defensin Lc-def (*Lens culinaris* defensin) is highly capable of inhibiting microbial growth of both Gram positive and Gram negative bacteria (Finkina et al. 2008). Hammami and co-authors (2008) postulated that Lc-def affects lipid binding and lipid transport, respectively, hence inhibiting the growth of fungi through electrostatic interaction with anionic lipid components of fungal membranes (Shenkarev et al. 2014). The detailed mechanism of the antifungal activity remains to be further studied.

The low performance of split yellow peas could be explained by the presence of inhibiting factors such as the peptides defensin Psd1 and Psd2 (*Pisum sativum* defensin) (Tam et al. 2015). Psd1 and Psd2, isolated from the seed of pea *Pisum sativum* (Almeida et al. 2000), were reported to have high antimicrobial activity against some fungi and yeasts. Both Psds were found to be effective against the fungus *Aspergillus versicolor* and the mould *Neurospora crassa* but have little effect on the fungus *Fusarium moniliforme* and the yeast *Saccharomyces cerevisiae* (Almeida et al. 2000). The antibacterial activity was mediated through inhibition of the potassium ($K^+$) channel (Almeida et al. 2002) on the cell membrane, whereas the antifungal activity was found to be related to its high affinity with membranes containing glucosylceramides (GlcCer) or ergosterol (Erg) (Amaral et al. 2019; Gonçalves et al. 2012). Moreover, both studies (Amaral et al. 2019; Gonçalves et al. 2012) found that Psd1 adsorbed to the membrane at lower Kp, whereas at higher Kp the Psd1 could be partially inserted into the membrane. Since Lc-def and Psds are water-soluble, there is a high potential that they leached into the soaking water of split yellow peas, thus limiting the growth of *Lactobacillus*, despite the abundance of water soluble carbohydrates: 0.69 g/100 g (Huang et al. 2018).

An unexpected result was observed for yellow soybeans. Despite minimal amounts of dry matter and fermentable carbohydrates in particular (0.04 g/100 g) (Huang et al. 2018) *Lactobacillus* grew extremely well. This result could be attributed to minerals. The mineral profile of legume soaking water revealed significantly less sodium (35 vs. 46–170 mg/100 g dry matter) for soy than pulses (Chap. 3). Accounting for dry matter quantified by Huang and collaborators (2018), the sodium levels in the soaking water tested were the following: soy 0.09 mg/100 g, pulses 0.18–1.9 µg/100 g (green lentils and haricot beans, respectively). Sodium is known to inhibit *Lactobacillus* growth (Terpou et al. 2019). Therefore, it is possible that the lower level of sodium present in soy soaking water allowed further microbial growth. For example, an ingredient delivering five times more prebiotic carbohydrates (chickpeas, 0.19 vs. 0.04 g/100 g) (Huang et al. 2018) contained 12 times more sodium than the soy counterpart.

This results are fascinating as they reveal dynamics that are far more complicated than what previously thought. Water soluble carbohydrates, antimicrobial peptides and sodium are three major variables to consider when evaluating prebiotic ingredients and choosing the suitable food matrix for delivery of probiotic microorganisms.

### 4.3.6 Antimicrobial Activity

The MacConkey agar plates showed no inhibition zones of *E. coli* for all legume soaking water (data not shown), in contrast with the inhibition area observed around the known antibiotic discs (mention the name of antibiotic discs). Therefore, legume soaking water did not exert any antimicrobial activity against the pathogenic Gram negative *E. coli*.

Numerous legume nutrients are known to inhibit the growth of pathogenic bacteria such as *E. coli*: peptides (Amaral et al. 2019; Finkina et al. 2008; Gonçalves et al. 2012), phenolic compounds (Araya-Cloutier et al. 2018), saponins (Hassan et al. 2010) and phytic acid (Zhou et al. 2019). The soaking water tested were shown to contain low amounts of these nutrients. Specifically, legume soaking water contained 0.03–0.60 g/100 g of protein, 0.08–0.44 mg/g of phenolics, 1–3 mg/g of saponins (Huang et al. 2018) and 0.01–0.04 g/100 g of phytic acid (Chap. 3). Therefore, low antimicrobial activity was expected.

## 4.4 Conclusions

Legume soaking water, liquid and freeze-dried, showed enormous potential as texturizer and prebiotic ingredients. Their high protein solubility (near 100%) and slightly acidic pH (5.7–4.7) makes them suitable for several food applications. The legume of origin had a major impact on the functionalities. For example, foaming ability was relevant only for the water of split yellow peas (due to its higher protein content) and for that of yellow soybeans (higher foaming ability of soy protein and low sodium content): 19%. Excellent emulsifying ability was observed for peas (50 $m^2/g$ for a 2% solution) resulting from high levels of soluble protein (hydrophilic) and insoluble protein (hydrophobic). As for flours, soaking water of legumes were more affine to oil than water, with average OAC values of 2.1–2.7 g/g. Remarkably, most legume soaking water acted as prebiotic for the bacteria *Lactobacillus*. Differences among samples suggested complex dynamics, involving growth promoting nutrients (soluble carbohydrates) and growth inhibiting factors (antimicrobial peptides Lc-def, Psd1 and Psd2, and sodium). Beans and soybeans were the best prebiotics.

In closing, the soaking water of legumes offer numerous opportunities to be upcycled into food ingredients, ranging from texturizers to prebiotics. Split yellow peas and yellow soybeans were the most interesting ingredients. Further work on their food applications is encouraged for a sustainable food production.

**Acknowledgments** This book chapter was realized with the contribution of the teaching funds offered by Lincoln University for the research projects called "FOOD 399 – Research Placement" (Bachelor) and "FOOD 699 – Research Placement" (Taught Master). Anirudh Sounderrajan performed the analysis of protein solubility, while Jingnan Zhu and Silu Liu measured the pH.

## References


Almeida, M. S., Cabral, K. M., Zingali, R. B., & Kurtenbach, E. (2000). Characterization of two novel defense peptides from pea (*Pisum sativum*) seeds. *Archives of Biochemistry and Biophysics, 378*(2), 278–286.

Almeida, M. S., Cabral, K. M., Kurtenbach, E., Almeida, F. C., & Valente, A. P. (2002). Solution structure of *Pisum sativum* defensin 1 by high resolution NMR: Plant defensins, identical backbone with different mechanisms of action. *Journal of Molecular Biology, 315*(4), 749–757.

Amaral, V. S. G., Fernandes, C. M., Felício, M. R., Valle, A. S., Quintana, P. G., Almeida, C. C., Barreto-Bergter, E., Gonçalves, S., Santos, N. C., & Kurtenbach, E. (2019). Psd2 pea defensin shows a preference for mimetic membrane rafts enriched with glucosylceramide and ergosterol. *Biochimica et Biophysica Acta (BBA)-Biomembranes, 1861*(4), 713–728.

Araya-Cloutier, C., Vincken, J. P., van Ederen, R., den Besten, H. M., & Gruppen, H. (2018). Rapid membrane permeabilization of Listeria monocytogenes and Escherichia coli induced by antibacterial prenylated phenolic compounds from legumes. *Food Chemistry, 240*, 147–155.

Bordenave, N., Hamaker, B. R., & Ferruzzi, M. G. (2014). Nature and consequences of non-covalent interactions between flavonoids and macronutrients in foods. *Food & Function, 5*(1), 18–34.

Caballero, B., Finglas, P., & Toldrá, F. (2015). *Encyclopedia of food and health*. Academic: Oxford, England.

Cilla, A., Bosch, L., Barberá, R., & Alegría, A. (2018). Effect of processing on the bioaccessibility of bioactive compounds–a review focusing on carotenoids, minerals, ascorbic acid, tocopherols and polyphenols. *Journal of Food Composition and Analysis, 68*, 3–15.

Damian, J. J., Huo, S., & Serventi, L. (2018). Phytochemical content and emulsifying ability of pulses cooking water. *European Food Research and Technology, 244*(9), 1647–1655.

Dickinson, E. (2003). Hydrocolloids at interfaces and the influence on the properties of dispersed systems. *Food Hydrocolloids, 17*(1), 25–39.

Finkina, E. I., Shramova, E. I., Tagaev, A. A., & Ovchinnikova, T. V. (2008). A novel defensin from the lentil Lens culinaris seeds. *Biochemical and Biophysical Research Communications, 371*(4), 860–865.

Gibson, G. R., & Roberfroid, M. B. (1995). Dietary modulation of the human colonic microbiota: Introducing the concept of prebiotics. *The Journal of Nutrition, 125*(6), 1401–1412.

Gonçalves, S., Teixeira, A., Abade, J., de Medeiros, L. N., Kurtenbach, E., & Santos, N. C. (2012). Evaluation of the membrane lipid selectivity of the pea defensin Psd1. *Biochimica et Biophysica Acta (BBA)-Biomembranes, 1818*(5), 1420–1426.

Güçlü-Üstündağ, Ö., & Mazza, G. (2007). Saponins: Properties, applications and processing. *Critical Reviews in Food Science and Nutrition, 47*(3), 231–258.

Hammami, R., Ben Hamida, J., Vergoten, G., & Fliss, I. (2008). PhytAMP: A database dedicated to antimicrobial plant peptides. *Nucleic Acids Research, 37*(suppl_1), D963–D968.

Hassan, S. M., Byrd, J. A., Cartwright, A. L., & Bailey, C. A. (2010). Hemolytic and antimicrobial activities differ among saponin-rich extracts from guar, quillaja, yucca, and soybean. *Applied Biochemistry and Biotechnology, 162*(4), 1008–1017.

Huang, S., Liu, Y., Zhang, W., Dale, K. J., Liu, S., Zhu, J., & Serventi, L. (2018). Composition of legume soaking water and emulsifying properties in gluten-free bread. *Food Science and Technology International, 24*(3), 232–241.

Jarpa-Parra, M. (2018). Lentil protein: a review of functional properties and food application. An overview of lentil protein functionality. *International Journal of Food Science & Technology, 53*(4), 892–903.

Johnson, C. R., Thavarajah, P., Payne, S., Moore, J., & Ohm, J. B. (2015). Processing, cooking, and cooling affect prebiotic concentrations in lentil (*Lens culinaris Medikus*). *Journal of Food Composition and Analysis, 38*, 106–111.

Kanatt, S. R., Arjun, K., & Sharma, A. (2011). Antioxidant and antimicrobial activity of legume hulls. *Food Research International, 44*(10), 3182–3187.

Kaur, M., & Singh, N. (2005). Studies on functional, thermal and pasting properties of flours from different chickpea (*Cicer arietinum L.*) cultivars. *Food Chemistry, 91*(3), 403–411.

Klupšaitė, D., & Juodeikienė, G. (2015). Legume: Composition, protein extraction and functional properties. A review. *Chemical Technology, 66*(1), 5–12.

Kruger, N. J. (2009). The Bradford method for protein quantitation. In *The protein protocols handbook* (pp. 17–24). Totowa: Humana Press.

Lafarga, T., Álvarez, C., Villaró, S., Bobo, G., & Aguiló-Aguayo, I. (2019). Potential of pulse-derived proteins for developing novel vegan edible foams and emulsions. *International Journal of Food Science & Technology, 52*(2), 475–481.

Liu, S., Singh, M., Wayman, A., Chen, D., & Kenar, J. (2017). Evaluation of soybean–navy bean emulsions using different processing technologies. *Beverages, 3*(2), 23.

Lopez-Martinez, L. X., Leyva-Lopez, N., Gutierrez-Grijalva, E. P., & Heredia, J. B. (2017). Effect of cooking and germination on bioactive compounds in pulses and their health benefits. *Journal of Functional Foods, 38*, 624–634.

Makri, E., Papalamprou, E., & Doxastakis, G. (2005). Study of functional properties of seed storage proteins from indigenous European legume crops (lupin, pea, broad bean) in admixture with polysaccharides. *Food Hydrocolloids, 19*(3), 583–594.

Martinez-Villaluenga, C., Frias, J., & Vidal-Valverde, C. (2008). Alpha-galactosides: antinutritional factors or functional ingredients? *Critical Reviews in Food Science and Nutrition, 48*(4), 301–316.

McSwain, M., Johnson, C. R., Kumar, S., & Thavarajah, P. (2019). Pulses, global health, and sustainability: Future trends. In *Health benefits of pulses* (pp. 1–17). Cham: Springer.

McWatters, K. H., & Cherry, J. P. (1977). Emulsification, foaming and protein solubility properties of defatted soybean, peanut, field pea and pecan flours. *Journal of Food Science, 42*(6), 1444–1447.

Mills, E. C., Huang, L., Noel, T. R., Gunning, A. P., & Morris, V. J. (2001). Formation of thermally induced aggregates of the soya globulin β-conglycinin. *Biochimica et Biophysica Acta (BBA)-Protein Structure and Molecular Enzymology, 1547*(2), 339–350.

Mitra, S., & Dungan, S. R. (1997). Micellar properties of Quillaja saponin. 1. Effects of temperature, salt, and pH on solution properties. *Journal of Agricultural and Food Chemistry, 45*(5), 1587–1595.

Mitra, S., & Dungan, S. R. (2000). Micellar properties of quillaja saponin. 2. Effect of solubilized cholesterol on solution properties. *Colloids and Surfaces B: Biointerfaces, 17*(2), 117–133.

Mohanty, D., Misra, S., Mohapatra, S. & Sahu, P. S. (2018). Prebiotics and synbiotics: Recent concepts in nutrition. *Food Bioscience*. 152–160.

Mojica, L., & de Mejia, E. G. (2018). Legume bioactive peptides. In *Legumes* (pp. 106–128), CPI Group. Croydon (UK).

Moussou, N., Corzo-Martínez, M., Sanz, M. L., Zaidi, F., Montilla, A., & Villamiel, M. (2017). Assessment of Maillard reaction evolution, prebiotic carbohydrates, antioxidant activity and α-amylase inhibition in pulse flours. *Journal of Food Science and Technology, 54*(4), 890–900.

O'Kane, F. E., Vereijken, J. M., Gruppen, H., & Van Boekel, M. A. (2005). Gelation behavior of protein isolates extracted from 5 cultivars of *Pisum sativum L. Journal of Food Science, 70*(2), C132–C137.

Pearce, K. N., & Kinsella, J. E. (1978). Emulsifying properties of proteins: Evaluation of a turbidimetric technique. *Journal of Agricultural and Food Chemistry, 26*(3), 716–723.

Raikos, V., Campbell, L., & Euston, S. R. (2007). Effects of sucrose and sodium chloride on foaming properties of egg white proteins. *Food Research International, 40*(3), 347–355.

Rehal, J., Beniwal, V., & Gill, B. S. (2019). Physico-chemical, engineering and functional properties of two soybean cultivars. *Legume Research-An International Journal, 42*(1), 39–44.

Ribeiro, B. D., Alviano, D. S., Barreto, D. W., & Coelho, M. A. Z. (2013). Functional properties of saponins from sisal (*Agave sisalana*) and juá (*Ziziphus joazeiro*): Critical micellar concentration, antioxidant and antimicrobial activities. *Colloids and Surfaces A: Physicochemical and Engineering Aspects, 436*, 736–743.

Sathe, S. K., & Salunkhe, D. K. (1981). Functional properties of the great northern bean (*Phaseolus vulgaris L.*) proteins: Emulsion, foaming, viscosity, and gelation properties. *Journal of Food Science, 46*(1), 71–81.

Schley, P. D., & Field, C. J. (2002). The immune-enhancing effects of dietary fibres and prebiotics. *British Journal of Nutrition, 87*(S2), S221–S230.

Serventi, L., Wang, S., Zhu, J., Liu, S., & Fei, F. (2018). Cooking water of yellow soybeans as emulsifier in gluten-free crackers. *European Food Research and Technology, 244*(12), 2141–2148.

Shenkarev, Z. O., Gizatullina, A. K., Finkina, E. I., Alekseeva, E. A., Balandin, S. V., Mineev, K. S., Arseniev, A. S., & Ovchinnikova, T. V. (2014). Heterologous expression and solution structure of defensin from lentil Lens culinaris. *Biochemical and Biophysical Research Communications, 451*(2), 252–257.

Shevkani, K., Singh, N., Chen, Y., Kaur, A., & Yu, L. (2019). Pulse proteins: Secondary structure, functionality and applications. *Journal of Food Science and Technology, 56*, 1–12.

Sila, A., Bayar, N., Ghazala, I., Bougatef, A., Ellouz-Ghorbel, R., & Ellouz-Chaabouni, S. (2014). Water-soluble polysaccharides from agro-industrial by-products: Functional and biological properties. *International Journal of Biological Macromolecules, 69*, 236–243.

Singhal, A., Karaca, A. C., Tyler, R., & Nickerson, M. (2016). *Pulse proteins: From processing to structure-function relationships* (p. 55). Grain Legumes: Rijeka, Croatia.

Sitohy, M., & Osman, A. (2010). Antimicrobial activity of native and esterified legume proteins against Gram-negative and Gram-positive bacteria. *Food Chemistry, 120*(1), 66–73.

Siva, N., Thavarajah, P., Kumar, S., & Thavarajah, D. (2019). Prebiotic carbohydrates in pulse crops towards improve human health. *Frontiers in Nutrition, 6*, 38.

Sosulski, F. W., & McCurdy, A. R. (1987). Functionality of flours, protein fractions and isolates from field peas and faba bean. *Journal of Food Science, 52*(4), 1010–1014.

Stantiall, S. E., Dale, K. J., Calizo, F. S., & Serventi, L. (2018). Application of pulses cooking water as functional ingredients: The foaming and gelling abilities. *European Food Research and Technology, 244*(1), 97–104.

Sumnu, G., Sahin, S., Aydogdu, A. & Ozkahraman, B.C. (2016). Effects of legume flours on batter rheology and cake physical quality. In *III International conference on agricultural and food engineering 1152* (pp. 175–182).

Swennen, K., Courtin, C. M., & Delcour, J. A. (2006). Non-digestible oligosaccharides with prebiotic properties. *Critical Reviews in Food Science and Nutrition, 46*(6), 459–471.

Tam, J. P., Wang, S., Wong, K. H., & Tan, W. L. (2015). Antimicrobial peptides from plants. *Pharmaceuticals, 8*(4), 711–757.

Terpou, A., Papadaki, A., Lappa, I. K., Kachrimanidou, V., Bosnea, L. A., & Kopsahelis, N. (2019). Probiotics in food systems: Significance and emerging strategies towards improved viability and delivery of enhanced beneficial value. *Nutrients, 11*(7), 1591.

Tolstoguzov, V. B. (1997). Protein-polysaccharide interactions. *Food Science And Technology-New York-Marcel Dekker*, 171–198.

Wang, M. P., Chen, X. W., Guo, J., Yang, J., Wang, J. M., & Yang, X. Q. (2019). Stabilization of foam and emulsion by subcritical water-treated soy protein: Effect of aggregation state. *Food Hydrocolloids, 87*, 619–628.

Zayas, J. F. (1997). Foaming properties of proteins. In *Functionality of proteins in food* (pp. 260–309). Berlin/Heidelberg: Springer.

Zhou, Q. I., Zhao, Y. U., Dang, H., Tang, Y., & Zhang, B. (2019). Antibacterial effects of phytic acid against foodborne pathogens and investigation of its mode of action. *Journal of Food Protection, 82*(5), 826–833.

# Chapter 5
# Soaking Water Applications

**Luca Serventi, Jingnan Zhu, Hoi Tung Chiu, Mingyu Chen, Neha Nair, Jiaying Lin, and Sachin Deshmukh**

## 5.1 Introduction

Legume soaking water is an exciting opportunity for the food industry. Their content of oligosaccharides, fibre and protein (Huang et al. 2018) allows a diverse range of applications: texture improvers (emulsifiers, stabilisers and thickeners) and source of nutrition (minerals and phytochemicals in particular). This is a new area of research with vast potential for food manufacturers. At this stage, three food applications were found for legume soaking water: frozen dessert (namely, vegan, coconut-based ice cream), savoury bakery (gluten-free crackers) and savoury leavened bakery (gluten-free bread) (Fig. 5.1). While the bread study was part of a published paper (Bird et al. 2017), the other two projects are presented for the first time in this book chapter. The first section proposes soaking water of chickpeas and split yellow peas as thickeners for coconut-based frozen dessert (vegan ice cream). The second section unveils the antistaling abilities of soaking water from for pulses (haricot beans, chickpeas, green lentils and split yellow peas) in gluten-free crackers. Finally, a third section discusses the softening effect of chickpea soaking water on crumb hardness in gluten-free bread.

L. Serventi (✉) · J. Zhu · H. T. Chiu · M. Chen · N. Nair · J. Lin · S. Deshmukh
Department of Wine, Food and Molecular Biosciences, Faculty of Agriculture and Life Sciences, Lincoln University, Lincoln, Christchurch, New Zealand
e-mail: Luca.Serventi@lincoln.ac.nz



L. Serventi, *Upcycling Legume Water: from wastewater to food ingredients*,
https://doi.org/10.1007/978-3-030-42468-8_5

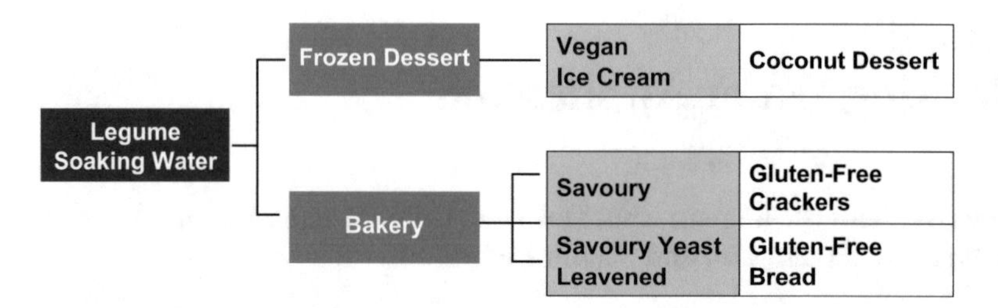


**Fig. 5.1** Applications of legume soaking water

## 5.2 Frozen Dessert

### *5.2.1 Plant-Based Ice Cream*

#### 5.2.1.1 Vegan Ice Cream

Ice cream is a frozen dessert consisting of a complex colloidal system of crystallised fat, water, air, minerals, sweeteners and proteins (Kaleda et al. 2018). The frozen state of ice cream includes ice crystals, air bubbles, fat globules and aggregates (Goff 2002). It is discrete and surrounded by the unfrozen state, which is continuous and includes sugars, proteins, salts, polysaccharides and water (Goff 2002). During manufacturing of ice cream, blending ingredients at freezing condition can cause fats at the air interface to absorb air bubbles, thus fat globule clusters are formed. However, the foaming structure is unstable, which can cause a collapse of volume (Goff 2002). Therefore, surface-active molecules such as some specific proteins, emulsifiers and surfactants, should be used for foaming stabilisation and formation to protect volume (Clarke 2015).

Typically, dairy products are used for ice cream making. Nonetheless, it might not be recommended for some consumer groups, for instance, those who are lactose intolerant and vegans (Kennedy 2018). Although there is no effective way to completely reduce the allergy (Mattar et al. 2012), dairy substitutes such as coconut products can be used to produce ice cream-like frozen desserts. The proteins found in coconut possess good emulsifying ability, but lower than that of milk protein (Naik et al. 2012). In addition, the protein content of coconut pulp used in coconut ice cream is low (0.8 g/100 g) (Iguttia et al. 2011). Therefore, coconut ice cream is often low in volume. Vegan ice cream is typically considered icy and coarse. This may be due to the difference in fat types (Choo et al. 2010) and protein content (Iguttia et al. 2011) from dairy milk. Furthermore, the lower of protein might decrease melting resistance and creaminess of the product owing to the lower emulsifying ability (Naik et al. 2012). Ingredients such as stabilizers and emulsifiers, the manufacturing process as well as the order of blending ingredients are crucial factors influencing the physical structure of ice cream (Granger et al. 2005).

Hydrocolloids, often called gums, are some of the common ingredients used as gelling agent and thickener, having the ability to enhance the viscosity of a liquid as a colloid mixture because the hydroxyl groups can enhance the affinity of hydrophilic substances and water to provide their capacity to gel (Saha and Bhattacharya 2010). Emulsion in the ice cream making process can reduce the surface tension, which can enhance desirable qualities in ice cream by enhancing whipping ability, increasing overrun capacity, decreasing whipping time, improving resistance to meltdown, inhibiting ice crystal growth, rising dryness and stiffness, imparting a smooth texture and a desirable slightly greasy mouthfeel, and enhancing product uniformity (Baer et al. 1997). According to scientific literature, legume soaking water exerts emulsifying ability and may soften food products like bread (Huang et al. 2018). Therefore, it could be used as an emulsifier to maintain air bubbles held by fat globules in ice cream. The objective of this work was to study the influence of soaking water from chickpeas and split yellow peas on the physicochemical characteristics and sensory quality of frozen dessert.

#### 5.2.1.2 Materials and Methods

Materials

Two kinds of legumes were studied in this project: garbanzo chickpeas (Kelley Bean Co., Scottsbluff, NE, USA) and split yellow peas (Cates, Ashburton, New Zealand). According to the preparation of legume soaking water from Bird et al. (2017), dry legumes were soaked in 3.3 times the weight of water for 16 hours at room temperature. Soaking yielded about 2.2 parts of water per part of dry legume. Soaking water were then boiled for 1 minute, cooled to room temperature and kept frozen at −18 °C until analyses.

Ice Cream Making

The amount of coconut cream, sugar, and vanilla extract was the same in the three recipes: 650 g, 120 g and 2 g, respectively. The amount of liquid varied based on the recipe: 228 g of tap water (control), 233 g of soaking water from split yellow peas (peas) or 229 g soaking water from chickpeas (chickpeas). Soaking water was whipped at speed 5 for 1 minute in a Delta Food Equipment mixer (500A, New Zealand) prior to the two main processing steps.

First, the liquid was blend with coconut cream (Nutribullet, Select, New Zealand) for 30 seconds on speed 1. Then, sugar and vanilla extract were added to the cream base and gently folded until all the ingredients were fully dissolved. Then, all ingredients were mixed in an ice cream machine (Cuisinart, New Zealand).

Instrumental Evaluation of Ice Cream

The weight of 100 ml of ingredients blend was recorded. After ice cream making, the weight of 100 ml of ice cream was measured, then overrun calculated as percentage of weight difference (mixture weight minus ice cream weight, divided by the ice cream weight, all multiplied times 100). The colour of each ice cream was measured with a colorimeter Minolta CR-210 (Japan). Hardness was determined by a TA.XT plus Texture Analyzer (Stable Micro Systems, United Kingdom) loaded with a 5 kg load cell and 5 mm cylinder probe. Samples of 60 ml of ice cream were immediately taken out of the ice cream machine, placed in a 100 ml graduated beaker, then tested at room temperature by single compression of 40% at a speed of 1.7 mm/s. Results were processed with a Texture Expert software. Finally, melting rate was determined on 100 g samples, placed on a sieve (6 holes/cm), inserted in a funnel covered graduated cylinder. The temperature was kept at 16 °C and the volume of ice cream melted was recorded every 10 minutes for 50 minutes.

Sensory Evaluation of Ice Cream

A group of 30 untrained panelists from Lincoln University (New Zealand) evaluated five samples: control, soaking chickpeas, soaking peas, cooking chickpeas and cooking peas. Samples were prepared for analysis 1 week before evaluation and stored at −10 °C until experiments. Ratings for colour, creaminess, sweetness and overall preference were performed on 9-point hedonic scales. All results are presented but discussion will be limited to a comparison among soaking water and control as this is focus of this section.

#### 5.2.1.3 Statistical Analysis

All analyses were performed in triplicate. Average and standard deviations of results were processed by Microsoft® Excel® 2016. The ANOVA one-way analysis ($\alpha = 0.05$) and Tukey's test were included in the statistical evaluation with Minitab®18.

Results and Discussion

#### 5.2.1.4 Instrumental Analysis

Legume soaking water is known to contain small amounts of protein: 0.08 and 0.60 g/100 g for chickpeas and split yellow peas, respectively (Huang et al. 2018). Based on these compositions, foaming ability could be expected from pea soaking water, thus increasing the volume of frozen desserts. Nonetheless, overrun measurements indicated loss of volume. The control recipe exhibited an overrun of 14%, significantly higher than chickpea soaking (1.9%) and pea soaking (5.0%). Coconut cream possess foaming and emulsifying abilities that allow for sufficient ice cream quality (Choo et al. 2010) thus explaining the small volume increase of 12.2%. Proteins found in chickpeas and peas have been proven to exhibit foaming and emulsifying abilities (Boye et al. 2010; Karaca et al. 2011). In addition, saponins and phenolic compounds can act as surfactants (Güçlü-Üstündağ and Mazza 2007). Finally, soluble polysaccharide found in legumes can stabilize emulsions (Dickinson 2018). The soaking water tested contained low amounts of polysaccharides, thus suggesting an insufficient air retention. It is possible that protein incorporated air but the lack of polysaccharides did not allow retention of such air. On the contrary, coconut cream contains polysaccharides (Góral et al. 2018) so the addition of solids from soaking water might have altered their network during ice cream making, thus explaining the loss of volume between the steps of blending (blender) and churning/cooling (ice cream machine).

Similarly, minor changes were detected in colour. Lightness did not significantly vary across products, while a∗ and b∗ components were higher in the chickpea recipe (Table 5.1). The seed coat of chickpeas contain tannins (Beninger and Hosfield 2003) which could be responsible for the higher red component of colour: a∗ 0.23 vs. ~0.00 (Table 5.1) Furthermore, phenolic compounds such as isoflavones (Xu et al. 2007) might have conferred yellow colour: b∗ 7.3 vs. ~6.0 (Table 5.1).

In terms of textural properties, all dessert recipes were soft, with hardness of about 60 g for control and chickpeas: 55.8 and 66.0 g, respectively. Interestingly, the pea dessert was almost 6 times harder: 316 g vs. 55.8 g. These numbers fell within an expected range for vegan ice cream/frozen dessert as they were well below 100 g. For example, a recent study on coconut milk-based ice cream revealed a hardness of 2.41 N, equal to 246 g (Góral et al. 2018), while another study reported a value of 848 g (Fuangpaiboon and Kijroongrojana 2015). In this specific study, desserts were analysed immediately after production, without frozen storage, thus explaining the lower hardness. Nonetheless, a noticeable hardening took place with the use of soaking water form yellow peas. Considering the high level of dry matter (1.89 g/100 g vs. 0.65 g/100 g of chickpeas) mainly consisting of soluble carbohydrates (Huang et al. 2018) it is hypothesized that soluble fibre absorbed water, thus resulting in hardening of the structure. Further evidence in support of this theory was obtained from the melting test. The control recipe started dripping after 5 minutes and lost an overall volume of 23 ml from the original volume of 60 ml (Fig. 5.2), in agreement with the literature (Choo et al. 2010; Góral et al. 2018). The soaking water from chickpeas did not have relevant impact, other than extending the first dripping time from 5 to 20 minutes, resulting in a comparable melting loss of 25 ml (Fig. 5.2). Remarkably, the soaking water from split yellow peas altered the melting properties of the frozen dessert. Firstly, no dripping was observed for 30 minutes (well above the 5 minutes of the control) (Fig. 5.2). Secondly, the final volumetric loss was only 18 ml. Finally, after the first drip, the rate of melting was comparable to that of the control. Both ingredients were recognized as good emulsifier, with

**Table 5.1** Colour of frozen desserts. Different letters represent significant difference ($p < 0.05$)

| Ice cream | L∗ | a∗ | b∗ |
|---|---|---|---|
| Control | 84.7 ± 1.4$^{a}$ | 0.03 ± 0.04$^{b}$ | 5.60 ± 0.04$^{b}$ |
| Chickpeas | 83.8 ± 0.3$^{a}$ | 0.23 ± 0.05$^{a}$ | 7.34 ± 0.13$^{a}$ |
| Peas | 83.0 ± 0.8$^{a}$ | −0.03 ± 0.03$^{b}$ | 6.29 ± 0.60$^{b}$ |

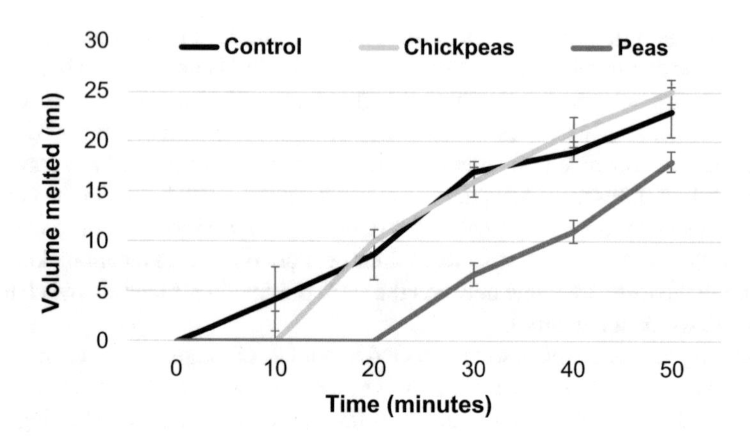


**Fig. 5.2** Melting profiles of the three recipes tested

emulsifying activity of about 50% (Huang et al. 2018). It is possible that a more homogeneous emulsification took place in the two legume-containing formulas, resulting in higher stability of the frozen structure, explaining later dripping. The application of emulsifiers in ice cream may lead to not only smaller air cells but also more uniform distribution of air cells throughout the interior structure of the ice cream, which finally results into final products with lower melting rate and higher textural quality (Goff and Hartel 2013; Innocente et al. 2009). Emulsifiers promote the partial coalescence of fat globules by displacing adsorbed proteins from the fat interface, therefore contributing to lower melting rate and better retention of shape. In fact, several studies have consistently shown that an increase in partial coalescence of fat results in decreasing melting of ice cream (Goff and Hartel 2013; Muse and Hartel 2004). Other factors to consider are sugars and fibre. Sugars have been shown to reduce melting rate by means of enhanced water absorption (Goff and Hartel 2013; Silva Junior and Lannes 2011). Pea soaking water contained high amounts of soluble carbohydrates, likely sugars such as sucrose, raffinose and stachyose. Therefore, the high sugar content of peas compared to chickpeas could have reduced the amount of free water, thus resulting in lesser freezing. Regarding fibre, partially soluble fibre such as basil seed gum, carrageenan, guar gum and locust seed gum are known to prevent ice cream melting due to their hydrocolloid properties (Adapa et al. 2000; BahramParvar and Goff 2013; Hussain et al. 2017; Laustsen 2011; Venugopal and Abhilash 2010). Specifically, these fibers stabilize ice cream structure and thicken it by absorbing water. These effects were particularly beneficial when hydrocolloids were used in combination with emulsifiers. It is possible that the insoluble fibre found in pea soaking water acted as a hydrocolloid, stabilizing the frozen structure. The presence of both emulsifying and thickening properties introduces the soaking water of split yellow peas as an excellent ice cream additive to improve texture and melting.

**Table 5.2** Sensory analysis of the frozen desserts

| Ice cream | Colour | Creaminess | Sweetness | Overall |
|---|---|---|---|---|
| Control | 6.0 ± 1.5[ab] | 6.9 ± 1.3[a] | 6.6 ± 1.7[a] | 6.8 ± 1.2[a] |
| Soaking chickpeas | 6.3 ± 1.7[ab] | 6.1 ± 2.0[a] | 6.1 ± 1.9[a] | 6.2 ± 1.7[ab] |
| Soaking split yellow peas | 6.5 ± 1.4[a] | 6.3 ± 1.7[a] | 6.5 ± 1.4[a] | 6.6 ± 1.1[a] |
| Cooking chickpeas | 5.7 ± 1.7[ab] | 6.1 ± 1.6[a] | 5.6 ± 1.6[a] | 5.7 ± 1.6[ab] |
| Cooking split yellow peas | 5.2 ± 1.7[b] | 5.8 ± 2.0[a] | 5.3 ± 1.9[a] | 5.6 ± 1.7[b] |

Different letters mean significant difference across treatments (PCW) ($p < 0.05$)
Different letters refer to statistically different samples

## Sensory Analysis

Replacement of tap water with legume water had minimal effect on the sensory quality of frozen desserts, with improvements in the pea recipe. This assessment was done as part of a larger study that involved cooking water of the same legumes (chickpeas and peas). Interestingly, wastewater of split yellow peas increased acceptability of colour as soaking (6.5 vs. 6.0), while decreased it when cooking (5.2 vs. 6.0) (Table 5.2). Instrumental analysis of colour did not depict differences between control and soaking peas, suggesting a role of non-pigments such as sugars. Sugars are known to provide brightness, colour and creamy appearance to ice cream (Cadena et al. 2012; Ozdemir et al. 2015). Considering the soluble carbohydrate content of pea soaking water (0.69 g/100 g) (Huang et al. 2018) it is possible that the sugars contained in it enhanced appearance. On the contrary, higher levels of soluble carbohydrates such as those found in pea cooking water (1.11 g/100 g) (Stantiall et al. 2018), along with double levels of phenolic pigments (0.6 vs 0.3 mg/g) (Damian et al. 2018; Huang et al. 2018) could have resulted in unpleasant, unbalanced appearance. Interestingly, creaminess was perceived similarly for all recipes (Table 5.2), despite differences in melting rate (Fig. 5.2). Sweetness perception was not affected by soaking water as expected due to their low levels of minerals, phenolics and saponins (Huang et al. 2018). Inversely, cooking water contain large fractions of bittering compounds such as calcium, phenolics and saponins, as well as salty (iron, potassium and sodium) (Damian et al. 2018) thus reducing sweetness perception. Overall, sensory acceptability of the frozen desserts formulated with soaking water was comparable to that of the control: 6.2 and 6.6 vs. 6.8 for chickpeas and peas, respectively (Table 5.2), ranging from like slightly to like moderately. Lower ratings were observed for the cooking water (5.7 and 5.6, respectively), mainly driven by lower sweetness.

## 5.3 Bakery

### *5.3.1 Gluten-Free Crackers*

#### 5.3.1.1 Emulsifiers and Antistaling Agents in Bakery Products

Emulsifiers are substances that stabilize emulsions or allow their production in the first place. They are a group of small molecules in relation to interfacial and surface physical chemistry, with have amphiphilic properties (the coexistence of both hydrophilic and lipophilic properties in the same molecule) (Norn 2015). Triglycerides (complex lipids), polyols (glycerol, sucrose and sorbitol), organic hydroxyacids (citric acid, lactic acid and more) are common classes of food emulsifiers (Norn 2015). They may facilitate the development of a starch and lipid network during the production of baked goods, allowing for premium quality.

Antistaling agents also play an important role in bakery to prevent staling. Staling is a combination of chemical and physical changes in baked goods upon storage (Cauvain 1998). Changes in bakery products are linked to changes in the crystalline state of starch (retrogradation), which contributes to progressive firming of the crumb (Zobel and Kulp 1996) as well as losses of moisture and protein plasticization. As staling could affect the quality of baked products, and remains responsible for huge economic losses to the baking industry and consumer, today, several antistaling agents are used in the bakery industry. The antistaling agents include enzymes (like α-amylases, debranching enzymes, etc.), surface-active lipids (like lecithins and monoglycerides), shortening, carbohydrate ingredients (like hydrocolloids and damaged and modified starch) and many more. They may reduce the rate of staling and inhibit moisture migration outside the products (Gray and Bemiller 2003). Antistaling agents may have direct or indirect effects on starch retrogradation, and direct influences include degradation of the amylose and/or amylopectin and a delay in the swelling and rupture of the starch granules, and indirect effect mainly include the diluting of starch (Purhagen et al. 2011).

Legumes as source of several water soluble nutrients: protein (albumins, etc.), carbohydrate (sugar, fibre), phenolics, saponins, etc. (Zhao et al. 2005; Klupšaitė and Juodeikienė 2015) that may act as emulsifiers and antistaling agents. For instance, the size, shape, amino acid composition and chemical structure of protein may influence the properties (like hydration, viscosity, emulsifying, foaming, etc) of food products (Klupšaitė and Juodeikienė 2015) and saponins have been widely used in food industry as surface active and foaming agents (Güçlü-Üstündağ and Mazza 2007). Rehinan and collaborators (2004) reported that peas, lentils and beans lost fractions of dietary fibre (neutral detergent fibre, acid detergent fibre, cellulose, hemicellulose and lignin) after soaking in tap water, which means they may dissolve into water. Therefore, this study tested the soaking water of four pulses (haricot beans, chickpeas, green lentils and split yellow peas) as emulsifiers and antistaling agents. Specifically, the goal was to evaluate whether the nutrients leached in these soaking water may act as antistaling agents to prevent hardening and moisture loss.

The matrix chosen was gluten-free crackers due to their high starch content (more prone to retrogradation) and flat structure (focus on texture, not volume).

#### 5.3.1.2 Materials and Methods

Preparation of the Soaking Water

Haricot beans (SUNVALLLEY FOODS, NZ), chickpeas (Cates GRAIN & SEED), whole green lentils (McKENZIE'S, NZ) and split yellow peas (Brown's Best) were soaked with a 1:3.3 weight ratio (dry legume:water) for 16 hours at room temperature. Sample were stored at −18 °C, then thawed at room temperature prior to experiments.

Cracker Manufacturing

Tap water (140 g) was mixed with maize corn starch (105 g), rice flour (105 g), canola oil (30 g), salt (4 g), and baking powder (4 g). Soaking water was introduced in full replacement of tap water, at a dose of 141 g. Ingredients were combined and kneaded for 5 minutes in a benchtop mixer (500A, Delta Food Equipment, Canada). Dough was rolled by hands with a rolling pin to a 4 mm thickness, pricked with a fork, then bake (MOFFAT, Australia) at 230 °C for 15 minutes. Crackers were packaged in plastic bags and stored at room temperature.

Pasting Properties

The pasting properties of the five dough were tested with the method described by Bird and collaborators (2017). Briefly, 11.88 g of soaking water were combined with 15.61 g of distilled water, 1.5 of corn starch and 1.5 g of rice flour maize starch flour in a Rapid Visco Analyzer (RVA Super 4, Newport Scientific, New Zealand), then heated from 50 to 95 °C and cooled back to 95 °C, for a total time of 13 minutes.

Cracker Quality

A Texture Analyser equipped with a 50 kg load cell and a 3-point bend fixture (TA. XT2, Stable Micro System, UK) was used to determine crackers' hardness on a single compression. About 2.5 of crackers were dried in the oven at 105 °C to quantify moisture content (AACC 2001). Colour readings were taken from three separate points on the surface of the crackers using a tristimulus colour analyser (Minolta Chroma Meter CR210, Minolta Camera Co., Japan). The illuminant C (CIE, standard, 6774 K) was used. Results were expressed as $L*$ (brightness), $a*$ (redness) and $b*$ (yellowness).

#### 5.3.1.3 Statistical Analysis

All experiments were performed in triplicate. Statistical differences in cracker and soaking water characteristics were determined by one-way analysis of variance (ANOVA) with formulation being the variable analysed and Tukey's comparison test ($p < 0.05$).

#### 5.3.1.4 Results and Discussion

Pasting Properties

Replacing tap water with legume soaking water resulted in significant changes to the pasting properties of gluten-free crackers. The dough developed with the soaking water of split yellow peas exhibited the highest peak viscosity (209 vs. 171 RVU) and breakdown viscosity (35 vs. 21 RVU) (Table 5.3). Peak viscosity reflects the maximum swelling of starch granules. This could be adversely affected by the presence of protein competing for water (Barak et al. 2013). Breakdown viscosity reflects the stability of the paste during cooking: the lower the breakdown viscosity the higher the stability of the starch-protein network (Barak et al. 2013). A previous study on the composition of legume soaking water revealed the presence of relevant amounts of water soluble carbohydrates (0.69 g/100 g) and protein (0.60 g.100 g) in the soaking water of split yellow peas (Huang et al. 2018). Soluble fibre and protein from peas have been associated with water absorption capacity (Lam et al. 2018; Wang et al. 2002), possibly explaining the increased viscosity of the dough. Moderate levels of insoluble fibre were found in this wastewater: 0.34 g/100 g (Huang et al. 2018). The insoluble fraction could have destabilized the protein network, thus weakening it as shown by the higher breakdown viscosity (Table 5.3).

On the contrary, chickpea soaking water resulted in the lowest setback viscosity: 44 vs. 59 RVU (Table 5.3). Setback value indicates the recovery of viscosity during cooling of the heated flour suspension (Barak et al. 2013), reflecting those staling dynamics that involve starch recrystallization. This result was in agreement with findings from a bread study where chickpea soaking water reduced breakdown viscosity (Bird et al. 2017). This difference may suggest that starch recrystallized to a

**Table 5.3** Pasting properties of the crackers' dough tested

| Cracker recipe | Peak viscosity (RVU) | Breakdown (RVU) | Setback (RVU) |
|---|---|---|---|
| Control | $172 \pm 4.3^{b}$ | $21 \pm 0.8^{b}$ | $59 \pm 4^{bc}$ |
| Beans | $195 \pm 3.1^{ab}$ | $28 \pm 0.5^{ab}$ | $57 \pm 3^{bc}$ |
| Chickpeas | $173 \pm 5.3^{b}$ | $25 \pm 2.6^{b}$ | $44 \pm 4^{c}$ |
| Green lentils | $186 \pm 17.2^{ab}$ | $26 \pm 5.7^{ab}$ | $73 \pm 15^{b}$ |
| Split yellow peas | $209 \pm 16.4^{a}$ | $35 \pm 5.2^{a}$ | $125 \pm 2^{a}$ |

The same letters in the same column means no significant differences ($p < 0.05$)
Different letters refer to statistically different samples

lower extent in the chickpea recipe. It has been proposed that the high ratio of soluble to insoluble carbohydrates in chickpea soaking water might have conferred it emulsifying ability (Huang et al. 2018). A possible outcome of this functionality is higher stability of the amorphous structure of starch, thus delaying its morphing into crystalline state. On the contrary, peas showed higher setback viscosity, nearly doubling that of the control dough: 125 vs. 59 RVU (Table 5.3). Peas are known to contain more sugars, namely oligosaccharides, than other legumes (Sánchez-Mata et al. 1998) and sugars tend to accelerate staling by subtracting water from starch and proteins (Schiraldi and Fessas 2001). Therefore, it is possible that the sugars in split yellow pea soaking water played a negative role during storage.

### Cracker Quality

Instrumental evaluation of gluten-free crackers revealed significant differences in colour. Lightness changed slightly in the legume crackers: 95.8–98.0 vs 96.7 of control (Table 5.4). The presence of pigments, likely phenolic compounds such as isoflavones and tannins, explained these changes. Chickpeas and green lentils resulted in higher, positive a∗ (4.3 and 2.1 vs. −11) (Table 5.4), reflecting more red component of colour. Furthermore, chickpea and lentil crackers were characterized by lower, negative b∗ (−8 and −3 vs. 27) (Table 5.4), indicating more blue component of colour. Chickpeas and lentils are known to contain more tannins than other pulses such as peas (Fratianni et al. 2014). Legume tannins typically confer light brown colour, but when heated at temperatures higher than 40 °C their chemical structure changes resulting in dark brown colour (Zhong et al. 2018). Darkening of tannins is compatible with the red and blue components of colour observed for chickpeas and lentils and with the baking conditions: 230 °C for 15 minutes. Higher b∗ value (36 vs. 28), thus more high yellow component, was observed for split yellow peas, as expected due to their yellow pigmentation induced by carotenoids (Padhi et al. 2017).

Textural analysis was extremely interesting. As expected, the control cracker hardened, from 20.9 to 39.5 kg, over the 2 day storage (Fig. 5.3a), basically doubling in hardness (Fig. 5.3b). This change was attributed to staling and it could be ascribed to one or more of these three different mechanisms: moisture loss, starch retrogradation and loss of protein plasticization (Gray and Bemiller 2003). Moisture

**Table 5.4** Colour profile of crackers

| Cracker recipe | L∗ | a∗ | b∗ |
|---|---|---|---|
| Control | $96.7 \pm 0.4^{b}$ | $-11 \pm 1^{c}$ | $27 \pm 2^{b}$ |
| Beans | $96.3 \pm 0.1^{c}$ | $-11 \pm 0^{c}$ | $28 \pm 1^{b}$ |
| Chickpeas | $98.0 \pm 0.2^{a}$ | $4.3 \pm 1.3^{a}$ | $-7.6 \pm 2.5^{d}$ |
| Green lentils | $97.7 \pm 0.1^{b}$ | $2.1 \pm 0.5^{b}$ | $-2.7 \pm 0.9^{c}$ |
| Split yellow peas | $95.8 \pm 0.1^{d}$ | $-14 \pm 1^{d}$ | $36 \pm 2^{a}$ |

The same letters in the same column means no significant differences ($p < 0.05$)
Different letters refer to statistically different samples

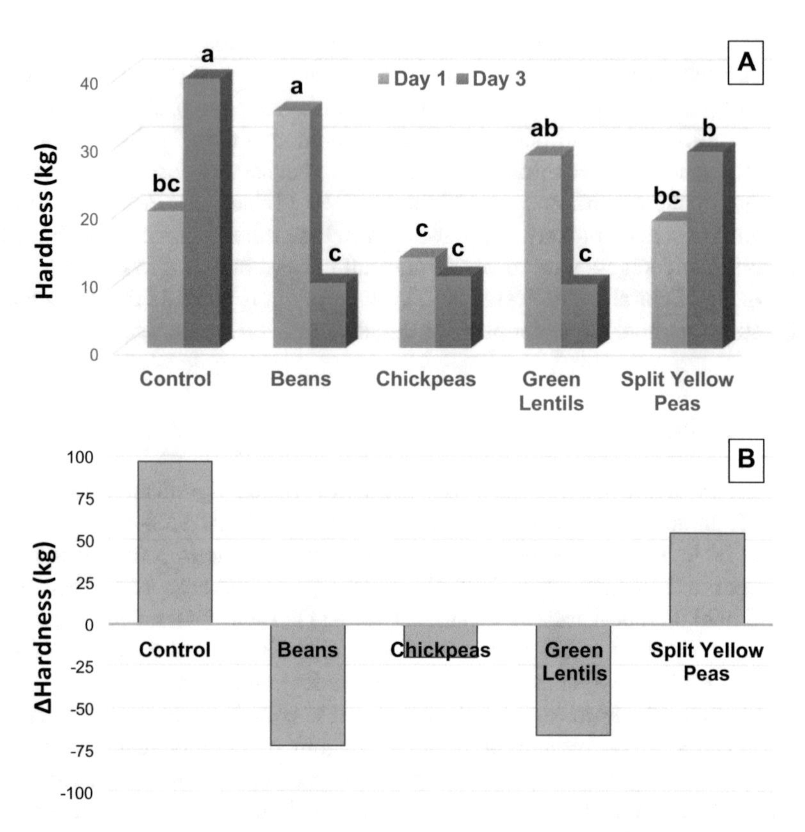


**Fig. 5.3** Hardness of crackers on days 1 and 3 of storage (**a**). Difference in cracker hardness (ΔHardness) between day 3 and day 1 of storage (**b**). Different letters refer to statistically significant difference ($p < 0.05$)

loss was not significant: 14.3 vs. 17 g/100 g (Fig. 5.3), thus ruling out this explanation. It was likely that staling of control crackers occurred due to starch recrystallisation and loss of protein elasticity. Bean crackers were harder (34.8 kg), while chickpea crackers were drastically softer (13.3 kg) (Fig. 5.3a). The emulsifying activity of these soaking water was determined to be high, at around 40-45% (Huang et al. 2018) and emulsifiers often soften bakery products by means of reduced surface tension among starch, protein, fat and air bubbles (Orthoefer 2008). Nonetheless, beans soaking water contains higher levels of solids, mostly insoluble fibre, than chickpea (Huang et al. 2018). These two characteristics might explain the changes in the fresh products. Storage revealed fascinating mechanisms. When calculating the difference in hardness after storage (ΔHardness), it was observed that only control and pea crackers hardened with storage: 100% and 55%, respectively (Fig. 5.3b). Chickpea crackers were fairly stable (−20%), while lentils and beans drastically softened by 67% and 73%, respectively (Fig. 5.3). Moisture content varied across fresh products, with lower values for beans, chickpeas and lentils (Fig. 5.4). These differences were probably due to the solids found in the soaking water, some of

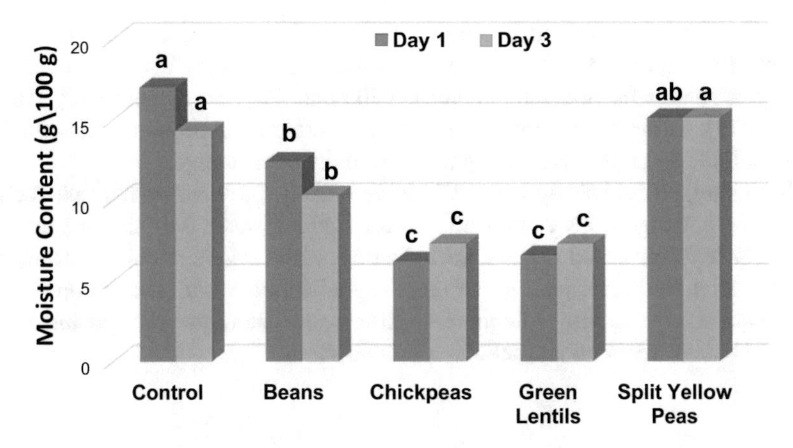


**Fig. 5.4** Moisture content of crackers on days 1 and 3 of storage. Different letters refer to statistically significant difference ($p < 0.05$)

which (perhaps insoluble fibre) might have disrupted the starch-protein network, thus causing higher water loss upon baking. Nonetheless, moisture content did not significantly change in any recipe, thus reasonably excluding moisture loss as cause of staling. Analysis of pasting properties did not show lower starch recrystallization in all legume recipes, consequently showing minor effects on starch structure. The last mechanism to be discussed is protein plasticization. It is possible that legume soaking water maintained high plasticization of the protein throughout storage. Two mechanisms are proposed: enhanced hydration due to water-binding soluble fibres; protein crosslinking with tannins. The soaking water of beans and peas contains high levels of water soluble carbohydrates (oligosaccharide and soluble fibre): 0.65–0.69 g/100 g (Huang et al. 2018). These macromolecules are known to absorb water and emulsify (Wang et al. 2002) thus could contribute to the hydration of protein, resulting in softening during storage. Nonetheless, pea crackers hardened over time. This phenomenon was expected due to their high setback viscosity measured in the RVA, likely resulting from sugars subtracting water from starch, which accelerated its transformation from amorphous to crystalline state. Macromolecules cannot explain the effect of chickpea and lentil soaking water, due to their low amount of solids: 0.65 and 0.39 g/100 g (Huang et al. 2018). A theory involving tannins might explain their strong antistaling effect. As shown by colour analysis (Table 5.4), beans and lentils crackers were likely sources of tannins. A study by Bordenave and collaborators (2014) has demonstrated the ability of certain phenolic compounds, namely kaempferol, kaempferol-3-glucoside; tiliroside and rutin, to act as emulsifiers. In addition, another study (Charlton et al. 2002) reported hydrophobic associations between the planar surfaces of aromatic rings of tannins and the hydrophobic sites of proteins such as prolyl residues, stabilised by hydrogen bonding between the hydrogen bond acceptor sites of proteins and the hydroxyl group of tannins. These complexes are soluble but, if further tannins are introduced, complexation extends to an insoluble form, causing protein precipitation. Similar mechanisms were described in another study, with tannins crosslinking to wheat proteins (Zhang and Do 2009). Bean and lentil soaking water both contained phenolic compounds in moderate amounts: ~300 and 400 μg/g, respectively (Huang et al. 2018). These concentrations might cause protein-tannin interaction within a water soluble form, thus enhancing their plasticity over storage.

In closing, textural changes in crackers were likely the results of multiple factors, due to the heterogeneous composition of the soaking water. Soluble fibre and tannins may have enhanced protein plasticization, while sugars may have decreased. Further, advanced investigations of these ingredients are warranted to understand the mechanisms occurring. Legume soaking water show potential as antistaling agent in bakery applications, acting on protein status.

### *5.3.2 Gluten-Free Bread*

Two studies applied legume soaking water in the development of bread, in both cases choosing gluten-free brad as matrix. The first study compared the soaking water of chickpea to other products and by-products of chickpea processing such as cooking water, flour and paste (Bird et al. 2017). Full replacement of tap water with chickpea soaking water affected the pasting properties by reducing setback viscosity from 63 to 51 RVU (Bird et al. 2017). As discussed in Sect. 5.2.1.2.2 (Cracker quality) the high ratio of soluble to insoluble carbohydrates could have conferred emulsifying properties to chickpea soaking water, thus preventing starch from recrystallizing. After fermentation and baking, bread loaf resulted significantly higher with chickpea soaking water than for the control (42 vs. 36 mm) without effects on the colour. Interestingly, crumb hardness was half for the chickpea-containing recipe than for the tap water counterpart: 2.2 vs. 4.3 kg (Bird et al. 2017). Furthermore, the structure improved. Scanning electron microscopy of crumb pores revealed less holes, lighter in colour (suggesting less depth) and with smoother crumb surface. It was speculated that the emulsifying property of chickpea soaking water softened the bread by retaining more air during baking, as shown by marginally higher loaf volume: 2.6 vs. 2.4 ml/g (Bird et al. 2017). Higher air incorporation is crucial in gluten-free bread to guarantee smooth pores surface and, consequently, sufficient moistness.

A comprehensive study investigated the soaking water of five legumes in the same gluten-free bread matrix (Huang et al. 2018). Pasting properties of the bread dough were deleteriously affected by the soaking water of yellow soybeans, resulting in higher breakdown and setback viscosity, thus weaker starch-protein network, more prone to retrogradation. The soy water used contained low levels of solids and no emulsifying properties so it is possible that it did not homogeneously interact with the other ingredients, destabilizing the starch-protein network. On the contrary, bread loaf was significantly softer when chickpea and pea soaking water were added, in agreement with findings by Bird ad collaborators (2017) on chickpea.

Unlike crackers, bread is manufactured by yeast fermentation. Therefore, the presence of sugars and antinutritional factors such as saponins must be considered. Soluble carbohydrates were present in relevant concentration for beans, chickpeas and peas, supporting similar emulsifying properties. Nonetheless, high levels of saponins (3 mg/g) were found in the soaking water of beans and peas (Huang et al. 2018), possibly inhibiting yeast metabolism (Güçlü-Üstündağ and Mazza 2007). Peas are known to contain more sugars than beans (Sánchez-Mata et al. 1998) so this might have overcome the negative effects of saponins on bread texture. Consequently, only chickpea and pea soaking water softened the crumb of gluten-free bread due to their soluble fibre and sugars. Bean emulsifying property was limited by their high amount of saponin content and low sugar content.

## 5.4 Conclusions

Legume soaking water have been applied in the development of confectionery and bakery products, specifically plant-based frozen desserts (coconut ice cream), gluten-free crackers and gluten-free bread. They showed potential for reducing ice cream melting (acting as hydrocolloids) and softening bread and crackers (antistaling effect).

The legume type affected their functionality. Soaking water of split yellow peas improved ice cream by reducing melting rate and volume, with enhanced visual appeal, due to its sugars and soluble fibre. Furthermore, the antistaling mechanism in crackers was ascribed to superior protein plasticization by soluble carbohydrates found in beans and chickpeas and tannins found in lentils. Bread manufacturing involves yeast fermentation of sugars, thus sugar and phytochemical contents must be taken into account: the first can promote loaf volume and softness while the latter might inhibit yeast metabolism.

Overall, legume soaking water showed great potential for textural improvement and stability of food, thus further trials are warranted for other food products.

**Acknowledgments** Data collection and discussion of this book chapter were possible with the financial support of Lincoln University to research projects at Bachelor level (FOOD 399 – Research Placement) and Master level (FOOD 699 – Research Placement). Further scientific discussion was implemented with the individual courses named “FOOD 698 – Research Essay”.

## References

AACC International. (2001). *Approved methods of analysis*. Method 44-19.01. Moisture—air-oven method, drying at 135° (11th edn). AACC International, St. Paul.

Adapa, S., Schmidt, K. A., Jeon, I. J., Herald, T. J., & Flores, R. A. (2000). Mechanisms of ice crystallization and recrystallization in ice cream: A review. *Food Reviews International, 16*(3), 259–271.

Baer, R. J., Wolkow, M. D., & Kasperson, K. M. (1997). Effect of emulsifiers on the body and texture of low fat ice cream. *Journal of Dairy Science, 80*(12), 3123–3132.

BahramParvar, M., & Goff, H. D. (2013). Basil seed gum as a novel stabilizer for structure formation and reduction of ice recrystallization in ice cream. *Dairy Science & Technology, 93*(3), 273–285.

Barak, S., Mudgil, D., & Khatkar, B. S. (2013). Relationship of gliadin and glutenin proteins with dough rheology, flour pasting and bread making performance of wheat varieties. *LWT-Food Science and Technology, 51*(1), 211–217.

Beninger, C. W., & Hosfield, G. L. (2003). Antioxidant activity of extracts, condensed tannin fractions, and pure flavonoids from Phaseolus vulgaris L. seed coat color genotypes. *Journal of Agricultural and Food Chemistry, 51*(27), 7879–7883.

Bird, L. G., Pilkington, C. L., Saputra, A., & Serventi, L. (2017). Products of chickpea processing as texture improvers in gluten-free bread. *Food Science and Technology International, 23*(8), 690–698.

Bordenave, N., Hamaker, B. R., & Ferruzzi, M. G. (2014). Nature and consequences of non-covalent interactions between flavonoids and macronutrients in foods. *Food & Function, 5*(1), 18–34.

Boye, J. I., Aksay, S., Roufik, S., Ribéreau, S., Mondor, M., Farnworth, E., & Rajamohamed, S. H. (2010). Comparison of the functional properties of pea, chickpea and lentil protein concentrates processed using ultrafiltration and isoelectric precipitation techniques. *Food Research International, 43*(2), 537–546.

Cadena, R. S., Cruz, A. G., Faria, J. A. F., & Bolini, H. M. A. (2012). Reduced fat and sugar vanilla ice creams: Sensory profiling and external preference mapping. *Journal of Dairy Science, 95*(9), 4842–4850.

Cauvain, S. P. (1998). Improving the control of staling in frozen bakery products. *Trends in Food Science & Technology, 9*(2), 56–61.

Charlton, A. J., Baxter, N. J., Khan, M. L., Moir, A. J., Haslam, E., Davies, A. P., & Williamson, M. P. (2002). Polyphenol/peptide binding and precipitation. *Journal of Agricultural and Food Chemistry, 50*(6), 1593–1601.

Choo, S. Y., Leong, S. K., & Henna Lu, F. S. (2010). Physicochemical and sensory properties of ice-cream formulated with virgin coconut oil. *Food Science and Technology International, 16*(6), 531–541.

Clarke, C. (2015). *The science of ice cream*. Royal Society of Chemistry, Bedford, UK.

Damian, J. J., Huo, S., & Serventi, L. (2018). Phytochemical content and emulsifying ability of pulses cooking water. *European Food Research and Technology, 244*(9), 1647–1655.

Dickinson, E. (2018). Hydrocolloids acting as emulsifying agents–How do they do it? *Food Hydrocolloids, 78*, 2–14.

Fratianni, F., Cardinale, F., Cozzolino, A., Granese, T., Albanese, D., Di Matteo, M., Zaccardelli, M., Coppola, R., & Nazzaro, F. (2014). Polyphenol composition and antioxidant activity of different grass pea (*Lathyrus sativus*), lentils (*Lens culinaris*), and chickpea (*Cicer arietinum*) ecotypes of the Campania region (Southern Italy). *Journal of Functional Foods, 7*, 551–557.

Fuangpaiboon, N., & Kijroongrojana, K. (2015). Qualities and sensory characteristics of coconut milk ice cream containing different low glycemic index (GI) sweetener blends. *International Food Research Journal, 22*(3), 1138–1147.

Goff, H. D. (2002). Formation and stabilisation of structure in ice-cream and related products. *Current Opinion in Colloid & Interface Science, 7*(5–6), 432–437.

Goff, H. D., & Hartel, R. W. (2013). *Ice cream*. Springer Science & Business Media, New York, NY, USA.

Góral, M., Kozłowicz, K., Pankiewicz, U., Góral, D., Kluza, F., & Wójtowicz, A. (2018). Impact of stabilizers on the freezing process, and physicochemical and organoleptic properties of coconut milk-based ice cream. *LWT, 92*, 516–522.

Granger, C., Leger, A., Barey, P., Langendorff, V., & Cansell, M. (2005). Influence of formulation on the structural networks in ice cream. *International Dairy Journal, 15*(3), 255–262.

Gray, J. A., & Bemiller, J. N. (2003). Bread staling: Molecular basis and control. *Comprehensive Reviews in Food Science and Food Safety, 2*(1), 1–21.

Güçlü-Üstündağ, Ö., & Mazza, G. (2007). Saponins: Properties, applications and processing. *Critical Reviews in Food Science and Nutrition, 47*(3), 231–258.

Huang, S., Liu, Y., Zhang, W., Dale, K. J., Liu, S., Zhu, J., & Serventi, L. (2018). Composition of legume soaking water and emulsifying properties in gluten-free bread. *Food Science and Technology International, 24*(3), 232–241.

Hussain, R., Singh, A., Vatankhah, H., & Ramaswamy, H. S. (2017). Effects of locust bean gum on the structural and rheological properties of resistant corn starch. *Journal of Food Science and Technology, 54*(3), 650–658.

Iguttia, A. M., Pereira, A. C., Fabiano, L., Silva, R. A., & Ribeiro, E. P. (2011). Substitution of ingredients by green coconut (*Cocos nucifera L*) pulp in ice cream formulation. *Procedia Food Science, 1*, 1610–1617.

Innocente, N., Biasutti, M., Venir, E., Spaziani, M., & Marchesini, G. (2009). Effect of high-pressure homogenization on droplet size distribution and rheological properties of ice cream mixes. *Journal of Dairy Science, 92*(5), 1864–1875.

Kaleda, A., Tsanev, R., Klesment, T., Vilu, R., & Laos, K. (2018). Ice cream structure modification by ice-binding proteins. *Food Chemistry, 246*, 164–171.

Karaca, A. C., Low, N., & Nickerson, M. (2011). Emulsifying properties of chickpea, faba bean, lentil and pea proteins produced by isoelectric precipitation and salt extraction. *Food Research International, 44*(9), 2742–2750.

Kennedy, S. (2018). Ice cream's healthy future. *Dairy Foods, 119*(3), 40–46.

Klupšaitė, D., & Juodeikienė, G. (2015). Legume: Composition, protein extraction and functional properties. A review. *Chemical Technology, 66*(1), 5–12.

Lam, A. C. Y., Can Karaca, A., Tyler, R. T., & Nickerson, M. T. (2018). Pea protein isolates: Structure, extraction, and functionality. *Food Reviews International, 34*(2), 126–147.

Laustsen, K. (2011). Dairy products and carrageenan: A perfect pairing. *Food Marketing & Technology, 5*, 7–9.

Mattar, R., de Campos Mazo, D. F., & Carrilho, F. J. (2012). Lactose intolerance: Diagnosis, genetic, and clinical factors. *Clinical and Experimental Gastroenterology, 5*, 113.

Muse, M. R., & Hartel, R. W. (2004). Ice cream structural elements that affect melting rate and hardness. *Journal of Dairy Science, 87*(1), 1–10.

Naik, A., Raghavendra, S. N., & Raghavarao, K. S. M. S. (2012). Production of coconut protein powder from coconut wet processing waste and its characterization. *Applied Biochemistry and Biotechnology, 167*(5), 1290–1302.

Norn, V. (2015). *Emulsifiers in food technology*. Wiley Blackwell, Hoboken, NJ, USA.

Orthoefer, F. (2008). Applications of emulsifiers in baked foods. In *Food emulsifiers and their applications* (pp. 263–284). New York: Springer.

Ozdemir, C., Arslaner, A., Ozdemir, S., & Allahyari, M. (2015). The production of ice cream using stevia as a sweetener. *Journal of Food Science and Technology, 52*(11), 7545–7548.

Padhi, E. M., Liu, R., Hernandez, M., Tsao, R., & Ramdath, D. D. (2017). Total polyphenol content, carotenoid, tocopherol and fatty acid composition of commonly consumed Canadian pulses and their contribution to antioxidant activity. *Journal of Functional Foods, 38*, 602–611.

Purhagen, J. K., Sjöö, M. E., & Eliasson, A. C. (2011). Starch affecting anti-staling agents and their function in freestanding and pan-baked bread. *Food Hydrocolloids, 25*(7), 1656–1666.

Rehinan, Z. U., Rashid, M., & Shah, W. H. (2004). Insoluble dietary fibre components of food legumes as affected by soaking and cooking processes. *Food Chemistry, 85*(2), 245–249.

Saha, D., & Bhattacharya, S. (2010). Hydrocolloids as thickening and gelling agents in food: A critical review. *Journal of Food Science and Technology, 47*(6), 587–597.

Sánchez-Mata, M. C., Peñuela-Teruel, M. J., Cámara-Hurtado, M., Díez-Marqués, C., & Torija-Isasa, M. E. (1998). Determination of mono-, di-, and oligosaccharides in legumes by high-performance liquid chromatography using an amino-bonded silica column. *Journal of Agricultural and Food Chemistry, 46*(9), 3648–3652.

Schiraldi, A., & Fessas, D. (2001). Mechanism of staling: An overview. In *Bread staling* (Vol. 1, pp. 1–17).

Silva Junior, E. D., & Lannes, S. C. D. S. (2011). Effect of different sweetener blends and fat types on ice cream properties. *Food Science and Technology, 31*(1), 217–220.

Stantiall, S. E., Dale, K. J., Calizo, F. S., & Serventi, L. (2018). Application of pulses cooking water as functional ingredients: the foaming and gelling abilities. *European Food Research and Technology, 244*(1), 97–104.

Venugopal, K. N., & Abhilash, M. (2010). Study of hydration kinetics and rheological behaviour of guar gum. *International Journal of Pharma Sciences and Research, 1*(1), 28–39.

Wang, J., Rosell, C. M., & de Barber, C. B. (2002). Effect of the addition of different fibres on wheat dough performance and bread quality. *Food Chemistry, 79*(2), 221–226.

Xu, B. J., Yuan, S. H., & Chang, S. K. C. (2007). Comparative analyses of phenolic composition, antioxidant capacity, and color of cool season legumes and other selected food legumes. *Journal of Food Science, 72*(2), S167–S177.

Zhang, X., & Do, M. D. (2009). Plasticization and crosslinking effects of acetone–formaldehyde and tannin resins on wheat protein-based natural polymers. *Carbohydrate Research, 344*(10), 1180–1189.

Zhao, Y. H., Manthey, F. A., Chang, S. K., Hou, H. J., & Yuan, S. H. (2005). Quality characteristics of spaghetti as affected by green and yellow pea, lentil, and chickpea flours. *Journal of Food Science, 70*(6), s371–s376.

Zhong, L., Fang, Z., Wahlqvist, M. L., Wu, G., Hodgson, J. M., & Johnson, S. K. (2018). Seed coats of pulses as a food ingredient: Characterization, processing, and applications. *Trends in Food Science & Technology, 80*, 35–42.

Zobel, H. F., & Kulp, K. (1996). The staling mechanism. In R. H. Hebeda & H. F. Zobel (Eds.), *Baked goods freshness,* Marcel Dekker, Inc., New York, USA.

# Chapter 6
# Cooking Water Composition

**Luca Serventi**

## 6.1 Boiling of Legumes

Legume processing consists of two main steps: soaking and cooking. Soaking provides softening of the seeds to allow for faster cooking. The soaked seeds can then be cooked in three different ways: boiling; canning; steaming.

Boiling is the easiest and most common way of cooking legumes. It is used, for example, in the production of tofu (Shurtleff and Aoyagi 2000). It can last anytime between 30 and 90 minutes, based on the legume type (Xu and Chang 2008).

Canning is used in the production of canned beans, chickpeas, lentils and peas (Schoeninger et al. 2017; Siddiq and Uebersax 2012). It is a 3-step process involving the following:

- Blanching at 80–85 °C for 4–30 minutes (Nleya et al. 2011; Parmar et al. 2016);
- Packaging in cans filled with brine (salt solution) (Nleya et al. 2011);
- Short cooking at 121 °C for about 15 minutes (Parmar et al. 2016).

Steaming is used in several food processes, such as the manufacturing of hummus, a chickpea-based spread (Xu et al. 2016). Steam cooking is a faster method, due to the application of autoclaves and pressure steaming technology. It is based on the high heat transfer capacity of water vapour and it reduces the cooking time to about 20 minutes (Rehman and Shah 2005).

Steaming produces minimum amounts of wastewater, while boiling and canning results in more significant levels of water waste. Canning wastewater (also known as brine) could be used by households and restaurants for small scale production of novel foods. On the contrary, boiling water could be used by a much larger audience: households, restaurants and food manufacturers. The aim of this book is

L. Serventi (✉)
Department of Wine, Food and Molecular Biosciences, Faculty of Agriculture and Life Sciences, Lincoln University, Lincoln, Christchurch, New Zealand
e-mail: Luca.Serventi@lincoln.ac.nz



L. Serventi, *Upcycling Legume Water: from wastewater to food ingredients*,
https://doi.org/10.1007/978-3-030-42468-8_6

recycling of industrial wastewater. Therefore, this chapter mainly focuses on boiling wastewater, describing its full composition: macronutrients, micronutrients and antinutrients. The information available on canning wastewater will be added to implement the discussion on the effect of cooking on nutrient leaching.

## 6.2 Macronutrients

Only two studies measured the loss of solids in the wastewater of canned legumes. The amount of solids in Aquafaba (name commonly adopted for canning water of chickpeas) was found to be 6–8 g/100 g (Buhl et al. 2019; Shim et al. 2018). This amount can be attributed to leaching of nutrients upon cooking, as well as during storage in the cans. The presence of protein was confirmed in the amount of 1.3–1.5 g/100 g of Aquafaba. Specifically, Buhl and collaborators (2019) found 13–14 g/kg of protein in Aquafaba, based on pH of the solution, while Shim and coauthors (2018) quantified 1.5 g/100 g. Minor differences can be attributed to variations in product quality based on the manufacturer and the batches produced.

The composition of legume wastewater after boiling was investigated in two studies (Serventi et al. 2018; Stantiall et al. 2018). These studies revealed dry matter to the extent of about 5 g/100 g in all wastewater. Solid content ranged from 4.4 g/100 g of split yellow peas to 5.6 g/100 g of yellow soybeans (Serventi et al. 2018; Stantiall et al. 2018). The only exception was represented by haricot beans, which only contained 3.3 g/100 g (Stantiall et al. 2018). This minor loss could be the result of a hard, smooth and thick husk. Overall, leaching in boiling water was comparable to what found for the canning water of chickpeas (from now referred to as Aquafaba) (Buhl et al. 2019; Shim et al. 2018). Solid leach in cooking water was slightly lower lower than that in the canning water (~5 vs. 6–8 g/100 g) possibly due to the effect of storage in the canned legumes, which resembles a prolonged soak.

The main fraction of nutrients in the cooking water of all legumes investigated was found to be insoluble fibre, ranging from moderate values (about 1/3 of solids from haricot beans) to high values (almost half of the solids, specifically 45%, for chickpeas and yellow soybeans) (Serventi et al. 2018; Stantiall et al. 2018). The difference could be explained by two factors: seed coat thickness, seed coat organization. Generally speaking, legumes have a similar architecture, consisting of the following three components: coat (representing 8–16% of the seed), the endosperm or cotyledon (80–90%) and the germ or embryo (1–3%) (Tiwari and Singh 2012). The seed coat represents the first protection from environmental factors and a supply media for nutrients during development, with outer and inner integument resulting from polymers such as cellulose, hemicellulose and lignin polymers (Tiwari and Singh 2012). Typically, thicker coats are found in lipid-rich legumes, such as soy and chickpeas (Tiwari and Singh 2012). Therefore, thicker coats equals higher amounts of insoluble fibre. It is not surprising that the two highest contents of insoluble fibre were found in the cooking water of yellow soybeans (2.46 g/100 g) (Serventi et al. 2018) and chickpeas (2.37 g/100 g) (Stantiall et al. 2018).

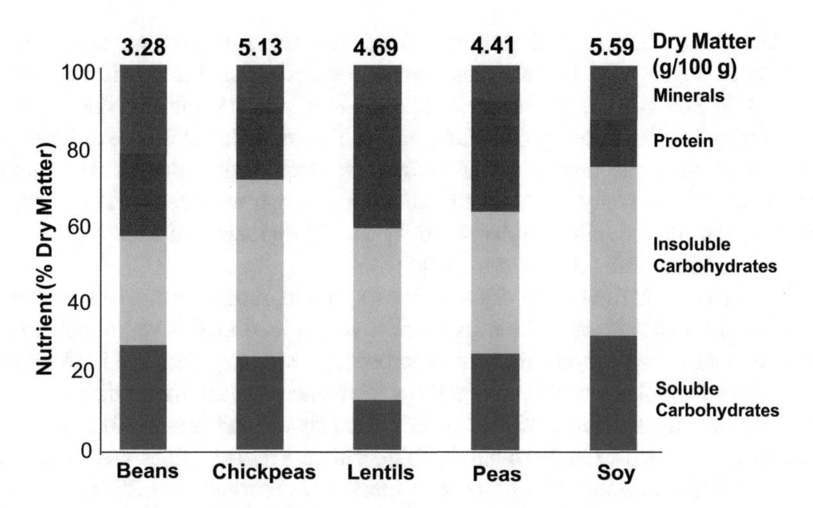


**Fig. 6.1** Nutrient distribution in legume cooking water (% dry matter). (Data from Stantiall and collaborators (2018) and Serventi and collaborators (2018))

Differences arise in the thickness and structural organization of the coat. Organisation is higher in peas and mung beans, while it is lower for chickpeas and soy. For example, navy beans have one of the highest content of dietary fibre (24–30 g/100 g) (Tiwari and Singh 2012) but only released small amounts in the wastewater: 0.93 g/100 g (Stantiall et al. 2018). This can be explained with the high structural organization of their seed coat, thus limiting nutrient leach. On the contrary, chickpeas contain less dietary fibre (around 17 g/100 g) but their coat structure is more unstable, rough and not homogenous (Tiwari and Singh 2012). Consequently, chickpeas released more insoluble fibre in the cooking water: 2.37 g/100 g (Stantiall et al. 2018). No starch was found in any of the cooking water investigated. Starch is stored in the endosperm, which is not typically exposed to water during boiling. Therefore, this result was expected.

On the contrary, soluble carbohydrates were found, representing both low molecular weight molecules (sugars) and high molecular weight molecules (soluble fibre) (Stantiall et al. 2018). The soluble fractions of carbohydrates varied significantly from cultivar to cultivar (Fig. 6.1). For instance, only 12% of the dry matter from lentil water was constituted by soluble carbohydrates, compared to almost 30% of the dry matter from beans, chickpeas, peas and soybeans (Fig. 6.1). These results are in agreement with the findings on legume soaking water (Chap. 3), where lentils soluble carbohydrates were even lower than other legumes: 10 vs. 15–35% (Serventi et al. 2018). For cooking water, the difference between lentils and other legumes was smaller and it could be due to two factors: sugars and soluble fibre. Previous studies have shown that lentils contain significantly less sugars but more soluble fibre than other legumes. For example, sugar content of lentil was found to be 40 g/kg dry matter versus 74 g/kg of chickpeas (Martín-Cabrejas et al. 2006). On the contrary, the content of soluble fibre was drastically higher for lentils than chickpeas: 27 vs 15 g/kg dry matter (Martín-Cabrejas et al. 2006). Possibly, the main soluble carbohydrates to leach during soaking were sugars (higher water solubility), while both sugars and soluble fibre were released upon boiling (more energy required to solubilize soluble fibre, having less water solubility than sugars). Considering that the fibre content was comparable among legumes, the difference in soluble carbohydrates among cooking water was less evident than for the soaking ones. In a previous study, a significant reduction in sugar was observed in chickpeas and cowpeas after cooking (Mubarak 2005) consisting mainly of sucrose, raffinose, stachyose and mullein (Ghribi et al. 2015).

The third largest fraction of dry matter was protein, ranging from 10% to 30% of the solids leached, for soy and lentils, respectively (Fig. 6.1). It is very interesting to observe that the highest protein source across legumes (soy) resulted in the lowest protein loss in cooking water: 0.68 g/100 g (Serventi et al. 2018). As discussed earlier, soybeans have a thick, smooth seed coat (Tiwari and Singh 2012) therefore limiting nutrient loss. On the contrary, higher than average values were found for peas and lentils: 1.27 and 1.51 g/100 g, respectively (Serventi et al. 2018). The peas studied were yellow split. The split nature exposed the starchy-proteinaceous endosperm to the boiling water, thus explaining a high nitrogen loss. On the other hand, the whole green lentils investigated were very thin, thus increasing the surface area exposed to the processing media, thus enhancing protein leaching.

Finally, no lipids were found, but ash were, in amounts varying from about 10% of split yellow peas to more than 20% for haricot beans (Fig. 6.1). Most minerals are found in the outer layer of seeds, thus explaining higher loss upon soaking than cooking (20–30% vs. 10–20%, respectively) (Huang et al. 2018; Serventi et al. 2018; Stantiall et al. 2018). Two cooking water were significantly richer in minerals than all others: haricot beans and yellow soybeans: 0.75 and 0.78 g/100 g, respectively (Serventi et al. 2018; Stantiall et al. 2018). Previous research has established that soybeans are the highest source of minerals across legumes: 4.8 vs. 2.8–3.6 g/100 g (de Almeida Costa et al. 2006; Redondo-Cuenca et al. 2007). Surprisingly, haricot beans released more minerals than other legumes, despite similar content. It is possible that the seed conformation and mineral distribution was different, allowing for such different results. Unlike soaking water, the cooking liquids contain five times more dry matter (on average 5 g/100 g vs. less than 1 g/100 g) (Huang et al. 2018; Serventi et al. 2018; Stantiall et al. 2018). This data showed that legume cooking water can be an excellent source of minerals, particularly interesting for the industry of nutritional supplements. The following section will detail the mineral profile of each sample and discuss the nutritional implications.

## 6.3 Micronutrients

### *6.3.1 Mineral Profile*

The mineral distribution of legume cooking water is summarised in Table 6.1. Potassium was by far the most abundant mineral leached in the cooking water of all legumes tested, reaching quantities as high as 6410 mg/100 g of dry matter in the

**Table 6.1** Mineral content of legume cooking water (mg/100 g dry matter)

| Mineral content of legume cooking water (mg/100 g dry matter) | Beans | Chickpeas | Lentils | Peas | Soy |
|---|---|---|---|---|---|
| Potassium (K) | 6410$^{a}$ | 3766$^{c}$ | 3600$^{d}$ | 2378$^{e}$ | 5915$^{b}$ |
| Phosphorous (P) | 1119$^{a}$ | 643$^{d}$ | 765$^{c}$ | 578$^{e}$ | 814$^{b}$ |
| Sulphur (S) | 562$^{b}$ | 409$^{c}$ | 642$^{a}$ | 418$^{c}$ | 315$^{d}$ |
| Magnesium (Mg) | 718$^{a}$ | 344$^{b}$ | 291$^{c}$ | 217$^{d}$ | 711$^{a}$ |
| Calcium (Ca) | 228$^{a}$ | 121$^{c}$ | 90.6$^{d}$ | 69.2$^{e}$ | 178$^{b}$ |
| Sodium (Na) | 22.0$^{c}$ | 89.7$^{a}$ | 16.0$^{d}$ | 21.9$^{c}$ | 23.7$^{b}$ |
| Iron (Fe) | 22.2$^{b}$ | 11.8$^{d}$ | 15.4$^{c}$ | 11.6$^{d}$ | 23.4$^{a}$ |
| Zinc (Zn) | 3.93$^{c}$ | 3.63$^{d}$ | 5.95$^{a}$ | 4.94$^{b}$ | 3.48$^{e}$ |
| Manganese (Mn) | 1.87$^{b}$ | 2.24$^{a}$ | 1.28$^{c}$ | 0.82$^{d}$ | 0.72$^{e}$ |
| Copper (Cu) | 1.81$^{d}$ | 2.00$^{c}$ | 3.68$^{a}$ | 2.09$^{b}$ | 1.98$^{c}$ |
| Molybdenum (Mo) | 0.27$^{b}$ | 0.81$^{a}$ | 0.24$^{c}$ | 0.10$^{d}$ | N/A |

Data from Damian and collaborators 2018 and Serventi and collaborators (2018)
Different letters refer to statistically significant difference ($p < 0.05$). The term N/A means not available

case of haricot beans (Fig. 6.1) equal to 210 mg/100 g of cooking water (Damian et al. 2018). When considering the nutritional relevance of bean cooking water, 210 mg provide 5.5% of the recommended daily intake for men (Ministry of Health New Zealand 2019). The nutritional contribution increases to 168% when considering the solid fraction only (dry matter), thus highlighting the opportunity of using legume cooking water for extraction of minerals. Other legumes released significantly less potassium (Table 6.1) despite similar content in the raw seed (Iqbal et al. 2006) probably due to a different localization of this mineral in their seed.

Equally relevant levels were determined for magnesium, iron, copper and molybdenum. Haricot beans and yellow soybean were excellent sources of magnesium, with about 700 mg/100 g dry matter, double the amount recommended daily (Ministry of Health New Zealand 2019). Lentils and peas contained lower levels, down to about a third of beans and soy (Table 6.1), thus suggesting structural differences. One of the most interesting results was iron. High levels of iron were detected in the samples studied. As for potassium, the best iron sources were beans and soy, which covered more than 100% of the RDI for both men and women (Table 6.1) (Ministry of Health New Zealand 2019). Iron levels were found to be comparable across the five legumes investigated (beans, chickpeas, peas, lentils and soy): 6.6–7.5 mg/100 g (Sandberg 2002) thus the higher losses in beans and soy could be attributed to the localization of this mineral in their seeds. Copper was present in high amounts, around 100% RDI, this time with minor differences across samples (Table 6.1). A staggering result was observed for molybdenum. Analysis of pulses (beans, chickpeas, lentils and peas) cooking water revealed high levels for beans, lentils and peas (0.10–0.27 mg/100 g) and an extremely high level for chickpeas (0.81 mg/100 g) (Table 6.1). To the best of our knowledge, no information is available on the molybdenum content in the cooking water from soy. In terms of human diet, 100 g of dry matter from chickpea cooking water could provide 24 times the

RDI (34 μg) (Ministry of Health New Zealand 2019). When considering chickpea cooking water as a liquid, similarly to Aquafaba from canning, the nutritional contribution to molybdenum was 0.04 mg/100 g (Damian et al. 2018), corresponding to 117% of the RDI. In other words, if chickpea cooking water were to be considered as food, a 100 g serve would provide the daily dose of molybdenum needed by an adult. Molybdenum is particularly abundant in legumes (Sardesai 1993) so it is expected that high levels might leach into the cooking water.

Other minerals were found in high amounts, but with intermediate nutritional significance, including the following: sulphur, calcium, sodium, zinc and manganese (Table 6.1). Results from the literature indicate that the cooking water of legumes can be a valuable source of minerals, especially potassium, phosphorous, magnesium, iron, copper and molybdenum.

### *6.3.2 Phytochemicals and Vitamins*

Boiling of food can cause loss of phytochemicals and vitamins. The most abundant phytochemicals in legumes are phenolic compounds and saponins. To the best of our knowledge, only two studies investigated the phytochemical loss upon boiling of legumes, while none analysed the effect on vitamins. The main class of phytochemicals leached in the cooking water was saponins. Results varied significantly, ranging from 6.4 mg/100 g of yellow soybeans (Serventi et al. 2018) to 14 mg/100 g of whole green lentils (Damian et al. 2018). As discussed above, saponins are typically bound to the protein fraction of seeds (Güçlü-Üstündağ and Mazza 2007). Consequently, higher saponins content were expected in the protein-rich samples. Results confirmed this hypothesis as Fig. 6.2 shows a strong correlation between saponin and protein content. Saponins are considered as antinutritive factors (Güçlü-Üstündağ and Mazza 2007). In the last decade they have been reevaluated in light of their ability to reduce the absorption of cholesterol (Francis et al. 2002) and glucose (Shi et al. 2004) as well as acting as anti-inflammatory and antioxidant (Singh et al. 2017). Therefore, moderate amounts of these compounds could improve human health. In this regard, the cooking water of beans and soy seemed reasonable candidates for such role, containing respectively 7.8 and 6.8 mg/100 g (Damian et al. 2018; Serventi et al. 2018) (Fig. 6.2). The other samples tested contained 10–14 mg/100 g (Damian et al. 2018) (Fig. 6.2), thus representing a rich source of saponins. If legume cooking water were to be considered as functional, liquid ingredient, beans and soy would represent moderate sources of saponins. Whereas, if legume cooking water were considered as matrix for nutrient extraction, chickpeas, lentils and peas would be the richest sources of saponins. Apart from quantitative discussion, qualitative evaluations must be made. Soy saponins, also known as soyasaponin, represent a diverse group of compounds, classified into four groups: A, B (the most abundant), E, and DDMP, based on the composition of their

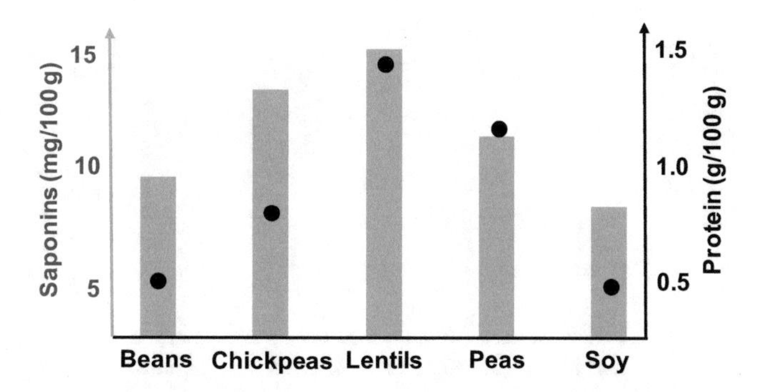


**Fig. 6.2** Saponin (bars) and protein (circles) contents of legume cooking water. (Data from Damian et al. (2018) and Serventi et al. (2018))

aglycone. Within each group, several types are present based on the profile of their sugar chain/s (Güçlü-Üstündağ and Mazza 2007). On the contrary, other legumes such as chickpeas and lentils only contain B-group saponins, and mostly one type: Bb (Singh et al. 2017; Serventi et al. 2013). Therefore, soy cooking water may provide several saponin groups and types in moderate amounts, while other legumes would not match the same level of diversity, but rather provide higher quantity of Bb saponin.

The other group of phytochemicals investigated was phenolics. Only two studies were found, focusing merely on pulses (Damian et al. 2018; Xu and Chang 2008). The levels found in the cooking water were low, spanning from 0.3 mg/100 g of haricot beans to about 0.6 mg/100 g of chickpeas, lentils and peas. Lentils are known to contain drastically more phenolics than other legumes: about 7 mg/g vs. 1.8, 1.7 and 2.6 mg/g of chickpeas, yellow peas and yellow soybeans, respectively (Xu et al. 2007). Nonetheless, similar values were found for all samples tested. Similarly to saponins, phenolics tend to be bound to protein (Ma et al. 2014). Unlike saponins, phenolics are heat labile (Xu et al. 2007), thus lower levels were recorded (Fig. 6.3). The study by Xu and Chang (2008) had shown that phenolic loss in the cooking water appears as soon as after 90 minutes. Therefore, the following minutes needed to complete the cooking may contribute to thermal degradation of the phytochemicals leached. Interestingly, similar levels of phenolics were found in the soaking water of the same legumes: 0.3 mg/100 g for beans and peas and 0.4 mg/100 g for lentils (Huang et al. 2018). Consequently, it can be concluded that legume soaking water are not a rich source of phenolics due to the heat-sensitive chemistry of this health-promoting compounds. Nonetheless, nutritionally significant levels (0.3–0.6 mg/100 g) are present, possibly providing health properties to these ingredients.

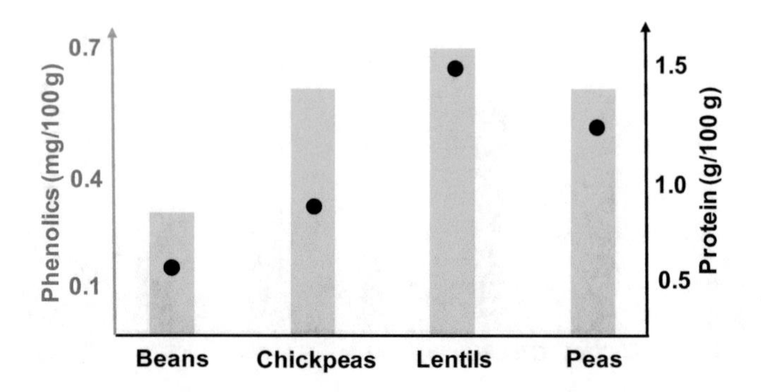


**Fig. 6.3** Phenolics (bars) and protein (circles) contents of pulses cooking water. (Data from Damian et al. (2018) and Xu and Chang (2008))

## 6.4 Antinutrients

### *6.4.1 Phytic Acid and Trypsin Inhibitor*

Phytic acid, classified as *myo*-inositol hexakisphosphate, is the main source of phosphorus storage in legume seeds in plant seeds and it carries negative effects on human health (Shi et al. 2018). It can be found more in legume soaking and cooking water due to the process of leaching (Ertaş and Bilgiçli 2014). Trypsin inhibitor represents a group of legume proteins with health-promoting activities (Clemente and del Carmen Arques 2014) as well as antinutritional effect of reduced protein absorption (Pesic et al. 2007). Cooking of legumes causes reduction of these antinutrients (Pisulewska and Pisulewski 2000). Therefore, leaching of phytic acid and trypsin inhibitor in the cooking water of legumes is possible. Hence, their quantification was performed.

### *6.4.2 Materials and Methods*

The amounts of antinutrients in legume cooking water were determined with the methods described in Chap. 3: phytic acid (Greiner et al. 2002; McKie and Mc Cleary 2016); trypsin inhibitors (Napoleão et al. 2013). Samples included haricot beans (Sun Valley Foods, New Zealand), chickpeas (Kelley Bean Co, NE, USA), green lentils (McKenzie's, Australia), split yellow peas (Cates, New Zealand) and yellow soybeans (YZ NON-GMO BEAN, Sunson, New Zealand).

### 6.4.3 Results and Discussion

A study on legume soaking water showed minimal leaching of phytic acid, at around 0.01–0.05 g/100 g (Chap. 3). This was expected based on the small difference in phytate content between raw and soaked legumes (Shi et al. 2018). On the contrary, boiling caused greater loss of these chemical, with values ranging from 0.01 to 0.05 g/100 g of pulses to 0.4 g/100 g of soy (Fig. 6.4). These numbers are very low when compared to the original content typically found in raw seeds: from 2.3 to 20 mg/g dry matter (Campos-Vega et al. 2010), equal to 0.23 to 2.0 mg/100 g, for lentils and soy, respectively. When considering this dynamic in perspective, about 20% loss of phytic acid was observed for lentils (0.04 g/100 g in the cooking water) and soy (0.42 g/100 g in the cooking water) (Fig. 6.4). The higher amount found in soy was ascribed to higher content in the raw seed (Campos-Vega et al. 2010) rather than to the seed geometry. The loss of dry matter was comparable across most legumes (about 5 g/100 g for chickpeas, lentils, peas and soy) (Serventi et al. 2018; Stantiall et al. 2018). Therefore, it was hypothesized that the aleurone exposure to processing water occurred to a similar extent, thus resulting in similar phytate leaching. When adding this to the soaking loss (about 2%) it can be deducted that only about 22% of phytic acid found in raw legumes is eliminated by traditional processing. These results showed that legume cooking water contains phytic acid in drastically lower levels than boiled legumes (about 4 times lower), thus not representing any concern for human nutrition when considered as liquid ingredient. If considered as dry ingredient, the sample would be concentrated by a factor of 20 (dry matter ~5 g/100 g) (Serventi et al. 2018; Stantiall et al. 2018) thus delivering values from 0.20 g/100 g (pulses) to 8.0 g/100 g (soy). A dry cooking water from pulses would carry similar phytate content as boiled pulses themselves, while the soy counterpart would represent a nutritional challenge if used in a large amounts.

On the contrary, lower dynamics were observed for trypsin inhibitors. Soaking had a low impact on trypsin levels: 0.03–0.10 TUI/mg in pulses and 3.0 TUI/mg in soy soaking water (Chap. 3). Interestingly, boiling water contained lower amounts than their soaking counterparts: from 0.06 TUI/mg of chickpeas to 0.13 TUI/mg of lentils (Fig. 6.4). Studies on the effect of boiling on legumes' antinutrients showed

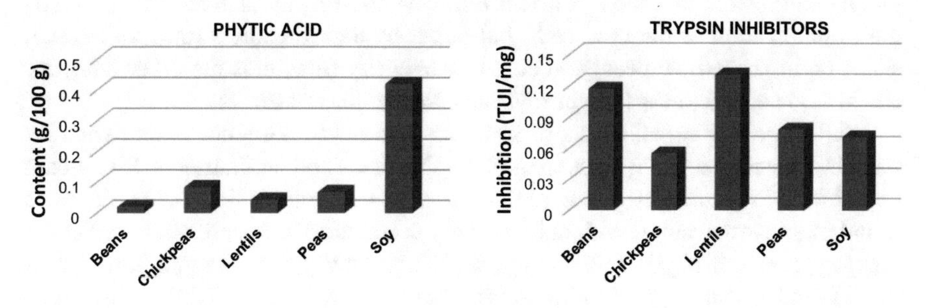


**Fig. 6.4** Contents of phytic acid (left) and trypsin inhibitors (right) of legume cooking water

that the majority of trypsin inhibitor was inactivated by the boiling process. Specifically, considering an average boiling time of 90 minutes, the inactivation yield of trypsin inhibitor was 63–100% for beans, 100% for chickpeas, 81–88% for lentils, 38–100% for peas and 96–100% for soy (Avilés-Gaxiola et al. 2018; Baker and Mustakas 1973). One hypothesis was loss of this protein in the boiling water but results of this study rejected such hypothesis (Fig. 6.4). An explanation can be found in the high heat sensitivity of trypsin inhibitor, regardless of the substrate (pulses or other legumes). For example, processing navy beans (*Phaseolus vulgaris L.*) in water bath at 90 °C reduced trypsin activity by 90%, comparably to only 5 minutes of microwave heating (Jourdan et al. 2007). Similarly, boiling soybeans for as little as 9 minutes degraded most of the trypsin inhibitor found in the raw seeds (Avilés-Gaxiola et al. 2018). These results suggest very low stability of the protein to heat treatments, likely explaining the lower levels found in cooking water than in soaking water. When considering cooking water as a dry powder, the amounts would be concentrated by a factor of about 20, therefore potentially rising to 1.1 TUI/mg for chickpeas and 3.5 TUI/mg for beans. Typical levels of trypsin inhibitors found in raw legumes are 8.1–16 TUI/mg for chickpeas and 16 TUI/mg for beans (Avilés-Gaxiola et al. 2018; Shi et al. 2018). Therefore, dry powder from legume cooking water could carry low amounts of trypsin inhibitors, which could easily be inactivated by physical processes (heat, extrusion, ultrasound, and ultrafiltration), chemical processes (acids, bases) and biological processes (germination and fermentation) (Avilés-Gaxiola et al. 2018).

## 6.5 Conclusions

Boiling caused significant loss of nutrients in the processing water. Most legumes released about 5 g of solids in each 100 g of cooking water. Canning resulted in even higher contents, likely due to prolonged soaking in the brine during cans storage. The main micronutrient identified was insoluble fibre, followed by soluble carbohydrates (oligosaccharides and soluble fibre), protein and minerals. The mineral profile revealed high levels of potassium, with nutritionally relevant amounts of potassium, phosphorous, magnesium, iron and copper. Extremely high levels of molybdenum were recorded. Nutrient loss was not directly affected by the proximate composition of the raw seeds, but rather by their thickness (thin for lentils), shape (split for yellow peas) and coat homogeneity (low, thus more leaching, for chickpeas) and thickness (low in soybeans, thus high release).

While leaching explained most of macro- and micro-nutrients, other dynamics must be considered when discussing phytochemicals and antinutrients. Significant quantities of saponins (9–14 mg/g) and phenolic compounds (0.3–0.7 mg/g) were found in all wastewater tested, only partially correlating to the protein loss. On the contrary, levels of phytic acid and trypsin inhibitors were low; in fact, lower, than those found in soaking water. Phenolics and trypsin inhibitors are labile to heat, thus boiling loss was ascribed to thermal degradation and did not contribute to the nutritional profile of the wastewater.

In closing, legume cooking water represent an important source of oligosaccharides, soluble fibre, insoluble fibre, protein, minerals and saponins. Potential applications could be nutrient extraction, texturizers (emulsifiers, foaming and thickening), prebiotics and enzyme extraction. These possibilities will be discussed in the following chapter: Chapter 7: Cooking Water Functional Properties.

**Acknowledgments** The author acknowledges Kaviya Sathyanarayanan for quantifying phytic acid and Lirisha Vinola Dsouza for analysing trypsin inhibitor, both funded by the courses "FOOD 699 – Research Placement". Further acknowledgments go to Letitia Stipkovits for planning their experimental design.

## References

Avilés-Gaxiola, S., Chuck-Hernández, C., & Serna Saldivar, S. O. (2018). Inactivation methods of trypsin inhibitor in legumes: A review. *Journal of Food Science, 83*(1), 17–29.

Baker, E. C., & Mustakas, G. C. (1973). Heat inactivation of trypsin inhibitor, lipoxygenase and urease in soybeans: Effect of acid and base additives. *Journal of the American Oil Chemists' Society, 50*(5), 137–141.

Buhl, T. F., Christensen, C. H., & Hammershøj, M. (2019). Aquafaba as an egg white substitute in food foams and emulsions: Protein composition and functional behavior. *Food Hydrocolloids, 96*, 354–364.

Campos-Vega, R., Loarca-Piña, G., & Oomah, B. D. (2010). Minor components of pulses and their potential impact on human health. *Food Research International, 43*(2), 461–482.

Clemente, A., & del Carmen Arques, M. (2014). Bowman-Birk inhibitors from legumes as colorectal chemopreventive agents. *World journal of gastroenterology: WJG, 20*(30), 10305.

Damian, J. J., Huo, S., & Serventi, L. (2018). Phytochemical content and emulsifying ability of pulses cooking water. *European Food Research and Technology, 244*(9), 1647–1655.

de Almeida Costa, G. E., da Silva Queiroz-Monici, K., Reis, S. M. P. M., & de Oliveira, A. C. (2006). Chemical composition, dietary fibre and resistant starch contents of raw and cooked pea, common bean, chickpea and lentil legumes. *Food Chemistry, 94*(3), 327–330.

Ertaş, N., & Bilgiçli, N. (2014). Effect of different debittering processes on mineral and phytic acid content of lupin (*Lupinus albus L.*) seeds. *Journal of Food Science and Technology, 51*(11), 3348–3354.

Francis, G., Kerem, Z., Makkar, H. P., & Becker, K. (2002). The biological action of saponins in animal systems: A review. *British Journal of Nutrition, 88*(6), 587–605.

Ghribi, A. M., Sila, A., Gafsi, I. M., Blecker, C., Danthine, S., Attia, H., Bougatef, A., & Besbes, S. (2015). Structural, functional, and ACE inhibitory properties of water-soluble polysaccharides from chickpea flours. *International Journal of Biological Macromolecules, 75*, 276–282.

Greiner, R., Larsson Alminger, M., Carlsson, N. G., Muzquiz, M., Burbano, C., Cuadrado, C., Pedrosa, M. M., & Goyoaga, C. (2002). Pathway of dephosphorylation of myo-inositol hexakisphosphate by phytases of legume seeds. *Journal of Agricultural and Food Chemistry, 50*(23), 6865–6870.

Güçlü-Üstündağ, Ö., & Mazza, G. (2007). Saponins: Properties, applications and processing. *Critical Reviews in Food Science and Nutrition, 47*(3), 231–258.

Huang, S., Liu, Y., Zhang, W., Dale, K. J., Liu, S., Zhu, J., & Serventi, L. (2018). Composition of legume soaking water and emulsifying properties in gluten-free bread. *Food Science and Technology International, 24*(3), 232–241.

Iqbal, A., Khalil, I. A., Ateeq, N., & Khan, M. S. (2006). Nutritional quality of important food legumes. *Food Chemistry, 97*(2), 331–335.

Jourdan, G. A., Noreña, C. P., & Brandelli, A. (2007). Inactivation of trypsin inhibitor activity from Brazilian varieties of beans (*Phaseolus vulgaris L.*). *Food Science and Technology International, 13*(3), 195–198.

Ma, W., Guo, A., Zhang, Y., Wang, H., Liu, Y., & Li, H. (2014). A review on astringency and bitterness perception of tannins in wine. *Trends in Food Science & Technology, 40*(1), 6–19.

Martín-Cabrejas, M. A., Aguilera, Y., Benítez, V., Mollá, E., López-Andréu, F. J., & Esteban, R. M. (2006). Effect of industrial dehydration on the soluble carbohydrates and dietary fiber fractions in legumes. *Journal of Agricultural and Food Chemistry, 54*(20), 7652–7657.

McKie, V. A., & Mc Cleary, B. V. (2016). A novel and rapid colorimetric method for measuring total phosphorus and phytic acid in foods and animal feeds. *Journal of AOAC International, 99*(3), 738–743.

Ministry of Health New Zealand. Nutrient Reference Values. URL: https://www.nrv.gov.au/nutrients. Accessed on 01 Aug 2019.

Mubarak, A. E. (2005). Nutritional composition and antinutritional factors of mung bean seeds (*Phaseolus aureus*) as affected by some home traditional processes. *Food Chemistry, 89*(4), 489–495.

Napoleão, T. H., dos Santos-Filho, T. G., Pontual, E. V., da Silva Ferreira, R., Coelho, L. C. B. B., & Paiva, P. M. G. (2013). Affinity matrices of Cratylia mollis seed lectins for isolation of glycoproteins from complex protein mixtures. *Applied Biochemistry and Biotechnology, 171*(3), 744–755.

Nleya, T., Arganosa, G., Vandenberg, A., & Tyler, R. (2011). Genotype and environment effect on canning quality of kabuli chickpea. *Canadian Journal of Plant Science, 82*, 267–272.

Parmar, N., Singh, N., Kaur, A., Virdi, A., & Thakur, S. (2016). Effect of canning on color, protein and phenolic profile of grains from kidney bean, field pea and chickpea. *Food Research International, 89*(1), 526–532.

Pesic, M., Vucelic-Radovic, B., Barac, M., Stanojevic, S., & Nedovic, V. (2007). Influence of different genotypes on trypsin inhibitor levels and activity in soybeans. *Sensors, 7*(1), 67–74.

Pisulewska, E., & Pisulewski, P. M. (2000). Trypsin inhibitor activity of legume seeds (peas, chickling vetch, lentils, and soya beans) as affected by the technique of harvest. *Animal Feed Science and Technology, 86*(3–4), 261–265.

Redondo-Cuenca, A., Villanueva-Suárez, M. J., Rodríguez-Sevilla, M. D., & Mateos-Aparicio, I. (2007). Chemical composition and dietary fibre of yellow and green commercial soybeans (*Glycine max*). *Food Chemistry, 101*(3), 1216–1222.

Rehman, Z., & Shah, W. H. (2005). Thermal heat processing effects on antinutrients, protein and starch digestibility of food legumes. *Food Chemistry, 91*(2005), 327–331.

Sandberg, A. S. (2002). Bioavailability of minerals in legumes. *British Journal of Nutrition, 88*(S3), 281–285.

Sardesai, V. M. (1993). Molybdenum: An essential trace element. *Nutrition in Clinical Practice, 8*(6), 277–281.

Schoeninger, V., Coelho, S. R. M., & Bassinello, P. Z. (2017). Industrial processing of canned beans. *Ciência Rural, 47*(5), 1–9.

Serventi, L., Chitchumroonchokchai, C., Riedl, K. M., Kerem, Z., Berhow, M. A., Vodovotz, Y., ... & Failla, M. L. (2013). Saponins from soy and chickpea: Stability during beadmaking and in vitro bioaccessibility. *Journal of agricultural and food chemistry, 61*(27), 6703–6710.

Serventi, L., Wang, S., Zhu, J., Liu, S., & Fei, F. (2018). Cooking water of yellow soybeans as emulsifier in gluten-free crackers. *European Food Research and Technology, 244*(12), 2141–2148.

Shi, J., Arunasalam, K., Yeung, D., Kakuda, Y., Mittal, G., & Jiang, Y. (2004). Saponins from edible legumes: Chemistry, processing, and health benefits. *Journal of Medicinal Food, 7*(1), 67–78.

Shi, L., Arntfield, S. D., & Nickerson, M. (2018). Changes in levels of phytic acid, lectins and oxalates during soaking and cooking of Canadian pulses. *Food Research International, 107*, 660–668.

Shim, Y. Y., Mustafa, R., Shen, J., Ratanapariyanuch, K., & Reaney, M. J. (2018). Composition and properties of aquafaba: Water recovered from commercially canned chickpeas. *JoVE (Journal of Visualized Experiments), 132*, e56305.

Shurtleff, W., & Aoyagi, A. (2000). *Tofu & soymilk production: A craft and technical manual* (Vol. 2). Soyinfo Center.

Siddiq, M., & Uebersax, M. A. (2012). Dry beans and pulses production and consumption: An overview. In *Dry beans and pulses production, processing and nutrition* (pp. 3–22), Wiley, Hoboken, NJ, USA.

Singh, B., Singh, J. P., Singh, N., & Kaur, A. (2017). Saponins in pulses and their health promoting activities: A review. *Food Chemistry, 233*, 540–549.

Stantiall, S. E., Dale, K. J., Calizo, F. S., & Serventi, L. (2018). Application of pulses cooking water as functional ingredients: The foaming and gelling abilities. *European Food Research and Technology, 244*(1), 97–104.

Tiwari, B. K., & Singh, N. (2012). *Pulse chemistry and technology*. Royal Society of Chemistry, Cambridge, UK.

Xu, B. J., & Chang, S. K. C. (2007). A comparative study on phenolic profiles and antioxidant activities of legumes as affected by extraction solvents. *Journal of Food Science, 72*(2), S159–S166.

Xu, B., & Chang, S. K. C. (2008). Effect of soaking, boiling and steaming on total phenolic content and antioxidant activities of cool season food legumes. *Food Chemistry, 110*(1), 1–13.

Xu, G., Ye, X., Chen, J., & Liu, D. (2007). Effect of heat treatment on the phenolic compounds and antioxidant capacity of citrus peel extract. *Journal of Agricultural and Food Chemistry, 55*(2), 330–335.

Xu, Y., Cartier, A., Obielodan, M., Jordan, K., Hairston, T., Shannon, A., & Sismour, E. (2016). Nutritional and anti-nutritional composition, and in vitro protein digestibility of Kabuli chickpea (Cicer arietinum L.) as affected by differential processing methods. *Journal of Food Measurement and Characterization, 10*(3), 625–633.

# Chapter 7
# Cooking Water Functional Properties

**Luca Serventi, Congyi Gao, Mingyu Chen, and Venkata Chelikani**

## 7.1 Introduction

Legume ingredients such as flours, protein isolates, protein concentrates, soluble and insoluble fibre, individually and combined, are known to exert several functionalities. These ingredients can develop foams, emulsions and suspensions, affecting food texture (Aguilar et al. 2015; Foschia et al. 2017; Martens et al. 2017; Nilufer-Erdil et al. 2012; Tyler et al. 2017). Recently, consumer demand (Food Navigator USA 2019) and government programmes (European Commission 2019) have promoted sustainable practices to minimise the environmental impact of food processing. Thus, contemporary research is investigating ways to recycle by-products of food processing. One of the most interesting areas of study is the development of sustainable texturizers based on legume by-products, in both solid and liquid form. For example, the fibre- and starch-rich residue of soymilk production, okara, has shown excellent thickening abilities (Lian et al. 2019; Mateos-Aparicio et al. 2019; Ostermann-Porcel et al. 2017).

Okara was also shown to be a potential substrate for probiotic fermentation by lactobacilli (Vong and Liu 2019). Similarly, the wastewater from tofu production, called tofu whey, was successfully fermented into an alcoholic beverage (Chua et al. 2018: Chua and Liu 2019).

Only in the last 3 years, liquid by-products of legume processing (mainly cooking and canning water) have been considered and recycled as food ingredients (Bird et al. 2017; Damian et al. 2018; Lafarga et al. 2019; Mustafa et al. 2018; Serventi et al. 2018; Stantiall et al. 2018). These cooking water derive from the processing of several legumes, as illustrated in Fig. 7.1. Such liquids appear to be suspensions more than solutions (Fig. 7.1) and were characterised as emulsifiers, foaming agents

L. Serventi (✉) · C. Gao · M. Chen · V. Chelikani
Department of Wine, Food and Molecular Biosciences, Faculty of Agriculture and Life Sciences, Lincoln University, Lincoln, Christchurch, New Zealand
e-mail: Luca.Serventi@lincoln.ac.nz



L. Serventi, *Upcycling Legume Water: from wastewater to food ingredients*,
https://doi.org/10.1007/978-3-030-42468-8_7

**Fig. 7.1** Representative samples of legume cooking water. From left to right: haricot beans, chickpeas, green lentils, split yellow peas and yellow soybeans. (Photo by Luca Serventi, Ph.D.)

and gelling agents in seven studies (Buhl et al. 2019; Damian et al. 2018; Lafarga et al. 2019; Meurer et al. 2019; Mustafa et al. 2018; Serventi et al. 2018; Stantiall et al. 2018). Therefore, this chapter summarises and compares all research currently available on the functionality of legume wastewater as food ingredient: physicochemical properties (pH, density, viscosity and protein solubility) and physicochemical abilities (foaming, emulsifying and thickening). In addition, a new experimental section describes the microbiological properties of legume cooking water: pathogen inhibition and prebiotic activity.

## 7.2 Texturizer

### *7.2.1 Physicochemical Characteristics*

Protein solubility of legume cooking water was determined with the method described in Sect. 4.2.1. The vast majority of the protein fraction was found to be soluble, ranging from 86% of haricot beans to 100% of green lentils and yellow soybeans (Fig. 7.2). These results agreed with a recently published manuscript on chickpea cooking water. Meurer and collaborators (2019) prepared the ingredient with a similar method to Stantiall and co-authors (2018), determining a solubility of 863 μg/ml for chickpeas. This value is equivalent to 86%, similar to the 95% solubility recorded in this experiment. Pulses mainly consist of hydrophobic proteins: globulins (soluble in salt solutions) and prolamins (soluble in 70% aqueous solution of ethanol). The only water soluble fraction is represented by albumins (Jarpa-Parra 2018; Singhal et al. 2016). Therefore, it is likely that the protein fraction leached during boiling was highly soluble (possibly albumin) and the cooking environment (high temperature and high amount of water) did not affect their solubility. The low fraction of insoluble protein could be attributed to structural damage of the homogeneous legume seed (haricot beans) or to leach of inner storage proteins (typically

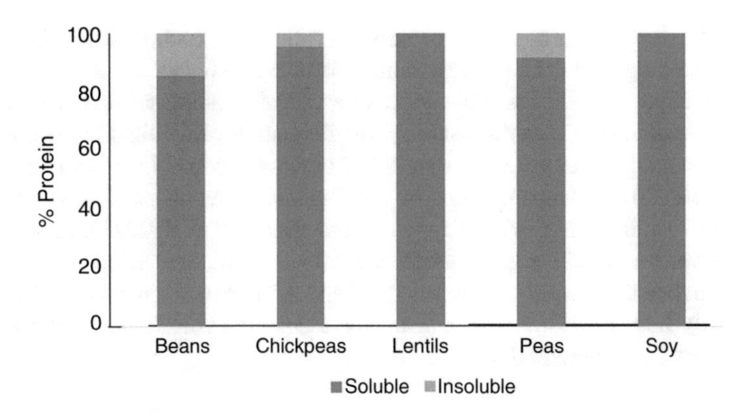


**Fig. 7.2** Protein solubility of legume cooking water

**Table 7.1** Reported values of pH, density and viscosity of legume cooking water

| | pH | References | Density (g/ml) | References | Viscosity (mPa∗s) | References |
|---|---|---|---|---|---|---|
| Haricot beans | **6.1** | Stantiall et al. (2018) | **1.02** | Stantiall et al. (2018) | **4.5** | Stantiall et al. (2018) |
| Chickpeas | **6.3** | Stantiall et al. (2018) | **1.01–1.03** | Meurer et al. (2019), Mustafa et al. (2018), and Stantiall et al. (2018) | **47–49** | Meurer et al. (2019) and Stantiall et al. (2018) |
| Green lentils | **6.5** | Stantiall et al. (2018) | **1.03** | Stantiall et al. (2018) | **25** | Stantiall et al. (2018) |
| Split yellow peas | **6.4** | Stantiall et al. (2018) | **1.02** | Stantiall et al. (2018) | **8.7** | Stantiall et al. (2018) |
| Yellow soybeans | **6.1** | Serventi et al. (2018) | **1.03** | Serventi et al. (2018) | **45** | Serventi et al. (2018) |

less water soluble) due to irregular surface of the seed coat (chickpeas) or to exposure of the endosperm to the water media (split yellow peas). Interestingly, ultrasound treatment increased protein solubility of the chickpea cooking water from 863 to 980 μg/ml. The highest solubility was reached upon application of ultrasounds at 20 kHz wave frequency and 34 W/cm$^2$ intensity for 30 minutes (Meurer et al. 2019).

Values of pH, density and viscosity of legume cooking water were reported in four peer-reviewed studies (Table 7.1). The pH of legume cooking water was found to be slightly acidic: from 6.1 of haricot beans and yellow soybeans to 6.5 of green lentils (Serventi et al. 2018; Stantiall et al. 2018). Density was found to be relatively low, with values ranging from 1.01 g/ml for chickpeas (Mustafa et al. 2018) to 1.03 g/ml for yellow soybeans (Serventi et al. 2018). Interestingly, three papers measured density of chickpea cooking water and observed different results: 1.01 g/ml (Mustafa et al. 2018), 1.02 g/ml (Stantiall et al. 2018) and 1.03 g/ml (Meurer et al. 2019). Processing had a major impact on the physicochemical properties of chickpea cooking water. The lowest value was obtained for the brine drained from canned chickpeas, while the other values described cooking water with different legume to water ratios. On the contrary, the two studies investigating viscosity of chickpea cooking water agreed on values of about 48 mPa∗s (Meurer et al. 2019; Stantiall et al. 2018). One other legume cooking water (yellow soybeans) was characterized by similarly high viscosity: 45 mPa∗s (Serventi et al. 2018). Intermediate viscosity was observed for green lentils (25 mPa∗s) while low viscosity characterised haricot beans and split yellow peas (4.8 and 8.7 mPa∗s, respectively) (Stantiall et al. 2018). Insoluble fibre was indicated as a potential contributor to the viscous behaviour of these ingredients

### *7.2.2 Foaming Properties*

Ever since the discovery that Aquafaba could be used as egg white replacer, studies on its foaming ability and stability took place. Five legume cooking water were tested in a study, reporting significant variations of foaming ability, from 38% of haricot beans to 97% of green lentils (Table 7.2), values comparable to flour solutions of similar concentration (Du et al. 2014). A linear correlation was observed between foaming ability and protein content (Stantiall et al. 2018). Proteins from pulses (beans, chickpeas, lentils and peas) and from soybeans are known to exert foaming ability (Boye et al. 2010; Matsumiya and Murray 2016; Toews and Wang 2013). In addition, the saponin content correlated well with foaming ability. For example, haricot beans contained 7.9 mg/g of saponins and developed a 38% foam fraction while, on the opposite end, green lentils contained 14 mg/g and developed

**Table 7.2** Foaming properties (capacity at time 0 and stability after 10 minutes) of legume cooking water

| | Foaming capacity (%) | References | Foaming stability (%) | References |
|---|---|---|---|---|
| Haricot beans | **38** | Stantiall et al. (2018) | | |
| Chickpeas | **58–548** | Lafarga et al. (2019), Meurer et al. (2019), Mustafa et al. (2018), and Stantiall et al. (2018) | **37–93** | Lafarga et al. (2019), Meurer et al. (2019), and Mustafa et al. (2018) |
| Green lentils | **97** | Stantiall et al. (2018) | | |
| Split yellow peas | **93** | Stantiall et al. (2018) | | |
| Yellow soybeans | **65** | Serventi et al. (2018) | | |

a 97% foam (Damian et al. 2018; Stantiall et al. 2018). Saponins commonly found in legumes are well established foaming agents and support foam stability by means of their surfactant abilities that reduce surface tension (Böttcher and Drusch 2017; Güçlü-Üstündağ and Mazza 2007).

Chickpea was the most studied legume, with four peer-reviewed publications. The foaming ability was measured as percentage of volume increase for the liquid or percentage of foam developed after whipping. Results varied drastically, from 58% to 548% (Table 7.2). Reasons for such variability resided in the preparation methods (water to legume ratio, cooking time) as well as in chemical and physical treatments of the liquid. To put these results in perspective, two studies compared chickpea cooking water against egg white, traditional foaming agent, observing lower values: 58 vs. 400% (Stantiall et al. 2018) and 259 vs. ~600% (Meurer et al. 2019). Another study proposed an evaluation of foaming properties of chickpea cooking water by hand shaking solutions. Results depicted a foam overrun of about 0.9 litre of foam per litre of solution, relevant but significantly lower than egg white: about 1.0 litre of foam per litre of solution (Buhl et al. 2019).

While untreated chickpea cooking water expressed foaming ability of 58–250%, higher values were recorded upon chemical and physical treatments such as low pH and application of ultrasound frequencies. Interestingly, the application of ultrasound frequency for 30 minutes raised the foam expansion of chickpea cooking water to 548%, nearly comparable to that of egg white (Meurer et al. 2019). Another study assessed the effect of pH on foaming ability and observed higher foaming properties at lower pH (Lafarga et al. 2019). Authors attributed these results to three mechanisms:

1. Ultrasounds modified the structure of chickpea proteins, enhancing their hydrophobicity, thus affinity for air bubbles (Meurer et al. 2019);
2. Ultrasounds homogenized the foam, resulting in more homogeneous distribution of air and protein particles (Meurer et al. 2019).
3. Lower pH increased protein net charges, improving their flexibility, thus reducing surface tension (Lafarga et al. 2019).

Regarding foaming stability, three studies monitored foams developed by whipping chickpea cooking water over time. Overall, stability of the chickpea foams was high, ranging from 37% to 93%, with the latter number occurring in most of the observations (Lafarga et al. 2019; Meurer et al. 2019; Mustafa et al. 2018). Ultrasound technology was shown to improve foaming stability due to the development of finer air bubbles (Meurer et al. 2019) while reduction of pH from 5.0 to 3.5 proved to stabilise foams by means of lower surface tension (Lafarga et al. 2019).

### *7.2.3 Emulsifying Properties*

Considering the presence of protein, soluble fibre and saponins in legume cooking water, emulsifying abilities are likely present in these ingredients. Consequently, six manuscripts investigated the ability of cooking water from five legumes to stabilise

**Table 7.3** Emulsifying properties of legume cooking water

| | Emulsifying activity index ($m^2/g$) | References | Emulsifying ability/ capacity (%) | References | Emulsifying stability | References |
|---|---|---|---|---|---|---|
| Haricot beans | **23** | Damian et al. (2018) | **46** | Damian et al. (2018) | | |
| Chickpeas | **15–39** | Buhl et al. (2019) and Damian et al. (2018) | **3.9–100** | Damian et al. (2018), Lafarga et al. (2019), Meurer et al. (2019), and Mustafa et al. (2018) | **23 minutes 0–76%** | Buhl et al. (2019) and Lafarga et al. (2019) |
| Green lentils | **47** | Damian et al. (2018) | **53** | Damian et al. (2018) | | |
| Split yellow peas | **16** | Damian et al. (2018) | **49** | Damian et al. (2018) | | |
| Yellow soybeans | **20** | Serventi et al. (2018) | **49** | Serventi et al. (2018) | | |

water and oil in emulsions. Five studies focused on chickpeas only, one on pulses (beans, chickpeas, lentils and peas) and one on soybeans. Three methods were chosen: Emulsifying Activity Index (EAI), Emulsifying Ability (EA) and Emulsifying Capacity (EC). The EAI is a spectrophotometric method that measures the extent to which an ingredient combines hydrophilic and hydrophobic components in a homogeneous matrix (Garcia-Vaquero et al. 2017; Pearce and Kinsella 1978). The EA and EC typically indicate the strength of a 50:50 water:oil emulsion (Chen et al. 2019; Jiang et al. 2009; Marefati et al. 2018). Furthermore, two studies monitored the stability of the emulsions over time.

An overview of legume cooking water can be obtained from two studies with similar sample preparation that analysed freeze-dried powders obtained from legume cooking water (Damian et al. 2018; Serventi et al. 2018). Moderate values of EAI were observed for haricot beans, split yellow peas and yellow soybeans: 23, 16 and 20 $m^2/g$, respectively (Table 7.3). Relevantly higher values of EAI were measured for chickpeas and green lentils: 39 and 47, respectively (Table 7.3). On the contrary, the EA test depicted similar strength for all emulsions tested, which were based on liquid samples of cooking water (Table 7.3). Emulsifying ability of legume ingredients is the result of one or more of the following factors:

- Soluble proteins (Ladjal-Ettoumi et al. 2016);
- Polysaccharides (Olorunsola et al. 2018; Ozturk and McClements 2016);

- Phospholipids (Sui et al. 2017);
- Saponins (Chung et al. 2017);
- Surfactant phenolics (Bordenave et al. 2014);
- pH (Ettoumi et al. 2016; Félix et al. 2019).

The cooking water of chickpeas and lentils contained similar amounts of dry matter (~5 g/100 g) and high amounts of protein: 0.95 and 1.51 g/100 g for chickpeas and lentils, respectively (Stantiall et al. 2018). As shown in Sect. 7.3.1, the protein fraction was largely soluble for both ingredients. Numerous studies reported high emulsifying activity for protein isolates from chickpeas and lentils (Felix et al. 2019; Jarpa-Parra 2018) as well as from haricot (navy) beans, peas and soy (Olorunsola et al. 2018; Tabtabaei et al. 2019; Xiong et al. 2018). In addition, the dry powders of cooking water studied consisted of comparable protein fractions (10–30%) with the pea water containing 1.27 g/100 g (Stantiall et al. 2018). Therefore, protein content alone did not explain the different functionalities among ingredients. Protein solubility was found to be lower for haricot beans: 86 vs. 100% (Sect. 7.3.1) thus possibly explaining its lower performance.

Furthermore, polysaccharides such as partially soluble fibre from legumes are known to emulsify (Olorunsola et al. 2018; Ozturk and McClements 2016) and smaller fractions were found for haricot beans and peas while a large fraction was found for chickpeas (Stantiall et al. 2018). It has been hypothesized that hemicellulose leached into chickpea cooking water (Stantiall et al. 2018) and hemicellulose from chickpeas was shown to act as emulsifier (Sanjeewa et al. 2010).

Phospholipids are excluded from the discussion of the emulsifying mechanisms of these ingredients since no lipids were detected in any of them (Serventi et al. 2018; Stantiall et al. 2018).

Phytochemicals known as saponins are likely to play a role in this activity. They have been shown to emulsify water and oil mixtures, developing finer particles than proteins and polysaccharides (Ozturk and McClements 2016). This mechanism was likely a cofactor in the high EAI for lentils and chickpeas, given their higher saponin content: 14 and 12 vs. 6–10 mg/g (Damian et al. 2018; Serventi et al. 2018). Interestingly, recent studies highlighted the contribution of certain phenolic compounds to emulsion development and stability. Specifically, kaempferol, kaempferol-3-glucoside, tiliroside and rutin exhibited emulsifying properties, particularly high for tiliroside and rutin (Bordenave et al. 2014). Both compounds are characterised by a sugar chain glycosidically linked to a flavonoid (tiliroside) or a flavonol (rutin) and are typically found in berries and citrusy fruits (Goto et al. 2012; Gullón et al. 2017). Lentils contain glycosides based on anthocyanins and condensed tannins: kaempferol glycoside, quercetin diglycoside, catechin and epicatechin (Zhang et al. 2018). Similarly, isoflavone glycosides were found in chickpeas (Xu et al. 2018). Consequently, conjugated phenolic compounds supported the emulsifying activity of the cooking water from chickpeas and lentils.

Finally, pH was considered. The ingredients tested presented similar pH values, from 6.1 to 6.5 (Table 7.1), thus excluding effects of pH to their functionalities. It can be concluded that the amount of soluble proteins, polysaccharides and saponins conferred different emulsifying activities to legume cooking water, while their combination guaranteed similar emulsifying ability (strength).

Chickpea was considered the most interesting ingredient by most researchers and their focus was on its properties. Results for EAI ranged from 15 to 39 $m^2/g$. The lowest value of EAI was measured for water drained from canned chickpeas (Buhl et al. 2019). The lower value for EA was found when chickpeas were boiled at a very high water to seed ratio (5 to 1) (Lafarga et al. 2019). It is likely that these preparations resulted in lower amounts of protein, polysaccharides and saponins in the ingredient, thus reducing their emulsifying potential. Stability of the emulsions was investigated by the same two studies described in the paragraph above. A 0% stability was recorded only when the ingredient was diluted and its pH reduced drastically (Lafarga et al. 2019). In regular, untreated conditions, the stability of the emulsions developed was determined to be high, up to 76% (Lafarga et al. 2019). Another study measured emulsions ability as the time needed for the EAI to drop significantly. Results showed that the emulsion developed with chickpea cooking water was drastically more stable than that made with pasteurised egg white powder: 23 vs. 9 minutes (Buhl et al. 2019). The same study also highlighted the emulsification potential of chickpea cooking water: EAI of 15 vs. 5 $m^2/g$ of egg white powder (Buhl et al. 2019). Positive results for chickpeas were correlated to smaller particle size of its emulsion (Buhl et al. 2019) which is known to support emulsion stability (Ozturk and McClements 2016). It is possible that proteins and polysaccharides developed a strong emulsion, while saponins stabilised it by means of reduced surface tension, thus smaller particle size. In addition, gelling abilities (water and oil absorption) might strengthen emulsions. Overall, the cooking water of chickpeas and lentils demonstrated to be excellent emulsifiers with wide potential for food applications.

### *7.2.4 Thickening Properties*

Aquafaba and other legume cooking water were successfully used as egg replacer in recipes for mousse, meringues, cakes and mayonnaise (Damian et al. 2018; Lafarga et al. 2019; Meurer et al. 2009; Mustafa et al. 2018; Stantiall et al. 2018). One of the mechanisms proposed was gelling ability. Gelling is typically the consequence of absorption of water and/or fats by protein and polysaccharides as demonstrated for eggs (Campbell et al. 2003), chickpeas and lentils (Aydemir and Yemenicioğlu 2013) as well as other legumes (Du et al. 2014). Therefore, water and oil absorption capacity of legume cooking water were investigated in two studies (Table 7.4). Water absorption capacity (WAC) was observed for three ingredients and in all cases with moderate values, ranging from about 1.5 g/g of chickpeas and yellow soybeans to 2.2 g/g of split yellow peas (Table 7.4). On the contrary, oil absorption capacity (OAC) was more pronounced and it occurred to similar extent in all five legume tested: from 2.68 g/g of yellow soybeans to 3.22 g/g of chickpeas (Table 7.4). These results were higher than those reported for legume flours, which indicated

**Table 7.4** Thickening properties of legume cooking water

| | Water absorption capacity (g/g) | Oil absorption capacity (g/g) | References |
|---|---|---|---|
| Haricot beans | **0.07** | **2.85** | Damian et al. (2018) |
| Chickpeas | **1.46** | **3.22** | Damian et al. (2018) |
| Green lentils | **0.13** | **2.71** | Damian et al. (2018) |
| Split yellow peas | **2.20** | **3.04** | Damian et al. (2018) |
| Yellow soybeans | **1.54** | **2.68** | Serventi et al. (2018) |

low WAC and OAC (approximately 1.3 g/g and 1.0 g/g, respectively) for navy beans, lentils and chickpeas (Du et al. 2014). High gelling ability was attributed to water binding polymers (proteins and polysaccharides) and to oil affinity of hydrophobic proteins (Aydemir and Yemenicioğlu 2013; Du et al. 2014). Soluble carbohydrates of low molecular weight (LMW) found in the cooking water were identified as the main responsible for their water absorption (Serventi et al. 2018; Stantiall et al. 2018). For example, the highest WAC was recorded for split yellow peas, which contained 1.0 g/100 g of low molecular weight carbohydrates (likely oligosaccharides) (Damian et al. 2018; Stantiall et al. 2018). On the contrary, virtually no water absorption (WAC 0.1 g/g) was observed for haricot beans and green lentils, which contained 0.7 and 0.5 g/100 g of LMW carbohydrates (Damian et al. 2018; Stantiall et al. 2018). These results suggest that the solids found in legume cooking water can absorb minor amounts of water.

Interesting results were obtained regarding oil absorption, with high values of approximately 3 g/g for all samples (Table 7.4). Low oil absorption was reported for legume flours: 0.9–1.2 g/g (Du et al. 2014). Oil absorption by legumes was mainly attributed to hydrophobic proteins, as confirmed by studies on the protein isolates of beans, chickpeas, lupine, lentils and soy (Foschia et al. 2017; Jarpa-Parra 2018). The results obtained for cooking water can only partially be explained by proteins because they represented 10–30% of the solid fraction (Serventi et al. 2018; Stantiall et al. 2018). The other major macromolecules are insoluble fibres. It is possible that polysaccharides with limited water solubility, such as hemicellulose and pectin, of which legumes are rich (Tosh and Yada 2010) enhanced oil affinity of these legume ingredients (Aravantinos-Zafiris et al. 1994). Insoluble fibre represented 35–50% of the solids in cooking water (Stantiall et al. 2018; Serventi et al. 2018) so this explanations seems plausible. These results show promising applications for legume cooking water as thickener in oil-based, fatty foods.

## 7.3 Microbiological Properties

### *7.3.1 Prebiotic Activity*

#### 7.3.1.1 Introduction

Legumes are rich in oligosaccharides that can support probiotics' growth (Chibbar et al. 2010). Common commercial legumes contain around 3.3–13.8% of soluble fibre (Subuola et al. 2012). Oligosaccharides include raffinose, α-galactosidase, stachyose and verbascose (de Fátima Viana et al. 2005). Apart from oligosaccharides, soluble fibre consists of β-glucans, galactomannan gums and pectins (Rodríguez et al. 2006).

Oligosaccharides are water-soluble so that they can be extracted by hydration and other wet processes (Tosh and Yada 2010). Kutoš and collaborators (2003) found many dietary fibres were lost during soaking, canning and cooking process, and canned beans lose the most amount of soluble fibres. Aquafaba attracts more interests because of its high nutrients and functionalities. Legume cooking water contain relevant amounts of soluble fibres (Serventi et al. 2018; Stantiall et al. 2018). Based on these findings, Aquafaba could be a source of prebiotic carbohydrates, to be used in the development of probiotic foods.

#### 7.3.1.2 Materials and Methods

Five legumes were studied: haricot beans (Sun Valley Foods, New Zealand), chickpeas (Kelley Bean Co, NE, USA), green lentils (McKenzie's, Australia), split yellow peas (Cates, New Zealand) and yellow soybeans (Sunson, New Zealand). Prebiotic properties of the legume cooking water were determined by adding 900 μl of soaking water samples to 100 μl of 0.8 OD *Lactobacillus acidophilus* cell suspension into 1.5 ml Eppendorf tubes. Samples were incubated at 37 °C for 24 hours anaerobically and after incubation, samples were serially diluted by a factor of $10^{-15}$ with 0.1% peptone water. 100 μl of these diluted samples were spread on the *Lactobacilli* MRS agar plates. The experiments were carried out in duplicates and CFU (colony forming units) were determined.

#### 7.3.1.3 Results and Discussion

The results of the total colonies of lactobacillus are shown in Table 7.5. The total colonies of peptone water and nutrient balance were 0 and 7 CFU, respectively (Table 7.5). As for samples, the lowest colony number was found for lentils cooking water: 4 CFU, therefore more than water but less than the nutrient broth (Table 7.5). Soybean cooking water allowed the growth of more colonies: 15 CFU (Table 7.5). At the same time, the total colonies of other three legume cooking water are too many to count which means the number of their total colonies more than 300 CFU.

**Table 7.5** Colony forming units (CFU) of *lactobacillus* in plates incubated with cooking water of five different legumes, dilutions $^{10-}6$, $^{10-}12$ and $10^{-15}$, peptone water and nutrient broth as control. The term TMC refers to too many to count

| | *Lactobacillus* (CFU) Dilutions | | |
|---|---|---|---|
| | $10^{-6}$ | $10^{-12}$ | $10^{-15}$ |
| Peptone water | 0 | 0 | 0 |
| Nutrient broth | TMC | 17 | 7 |
| Haricot beans | TMC | TMC | TMC |
| Chickpeas | TMC | TMC | TMC |
| Green lentils | TMC | TMC | 4 |
| Split yellow peas | TMC | TMC | TMC |
| Yellow soybeans | TMC | TMC | 15 |

There may be several reasons why lentil and soybean cooking water expressed lower prebiotic properties. During cooking, parts of soluble carbohydrates is transferred into the cooking water. Stantiall and collaborators (2018) and Serventi and collaborators (2018) observed that in legume cooking water, lentils have the lowest soluble carbohydrates content. However, soybean cooking water had the highest soluble carbohydrates, about 25% of dry matter (Stantiall et al. 2018). On the other hand, anti-nutrients also will influence the prebiotic property of legume cooking water. Anti-nutritional factors, such as trypsin inhibitors, phytic acid, and tannic acid, limit the use of protein and carbohydrates (Pal et al. 2017). Soybean cooking water has the highest phytic acid content: 0.4 g/100 g. Meanwhile, other cooking water only contained 0.01–0.05 g/100 g phytic acid (Chap. 6). Shi and co-authors (2018) reported that raw soybean seeds contain 23 mg/g phytic acid, drastically higher than peas and lentils (10 and 12 mg/g) so a higher phytate content in soybeans might have limited *lactobacilli* growth on soy cooking water. In contrast, haricot beans, chickpeas and split yellow peas significantly favoured the growth of this beneficial probiotic microorganisms, largely more effectively than the common nutrient broth.

## 7.3.2 *Antimicrobial Activity*

### 7.3.2.1 Introduction

Legume exerts high antimicrobial activity on both gram-positive and gram-negative bacteria, as well as moulds and fungi. Proteins, phytochemicals including saponins and phenolics, antinutrients like phytic acid, are known for the antimicrobial activity. Pina-Pérez and Ferrús Pérez (2018) elucidated that proteins and peptides from legumes reveal high microbial growth inhibition between 91% and 97%. Other legumes with antimicrobial activity comprises haricot beans, chickpeas, lentils, garden peas, soybeans, mung beans (*Vigna radiata*), pigeon peas (*Cajanus cajan*), and peanuts (*Arachis hypogeaea*). For instance, Both Kim and Rhee (2016) and Zhou and collaborators (2019) found that phytic acid at concentration between 0.2% and 0.16% (w/w) exerted excellent antimicrobial activity on both non-acid-adapted and acid-adapted *E. coli* O157:H7 strains. Furthermore, phytic acid is effective in destroying cell membranes of gram-positive bacteria (Zhou et al. 2019). Lipopolysaccharide (LPS) on the outer membrane (OM) of pathogenic bacteria (such as *E. coli* and *S. aureus*) protect the cell from antimicrobial agents or antibiotics (Nikaido 2003; Zhou et al. 2019) whereas phytic acid inhibits their growth through cell membrane damage by chelation of LPS. Both hydrogen ion and phytate ion can act on the OM-stabilizing divalent cations and disintegrate LPS on OM, through which the membrane integrity is disrupted. Furthermore, compared to the commonly used organic acids, phytic acid exerts greater bactericidal effect at same working concentration, particularly when in synergy with sodium chloride (Kim and Rhee 2016). In addition, phenolic compounds are known to inhibit bacterial growth (Araya-Cloutier et al. 2018), with saponins extending this activity to moulds and fungi (Hassan et al. 2010). Considering the high levels of protein, phenolics and saponins in legume cooking water (Buhl et al. 2019; Damian et al. 2018; Mustafa et al. 2018; Serventi et al. 2018; Stantiall et al. 2018) it was hypothesized that these ingredients might inhibit the growth of pathogenic bacteria such as *E. coli*.

#### 7.3.2.2 Materials and Methods

The antimicrobial activity of cooking water was tested on indicator organism using selective media MacConkey agar plates. Plates were spread with 100 μl 0.8 OD *E. coli* cell suspension and then the agar surface was dug out of the same size circular hole having diameter of 1.5 cm. Samples of each legume soaking water (1.5 ml) were poured into the wells, while an antibiotic disc was placed in the spare area for comparison. Experiments were carried out in duplicates and inhibition zones were measured.

#### 7.3.2.3 Results and Discussion

Results show no inhibition of *E. coli* (data not shown). Shehata and Marr (1971) reported that the concentration of glucose, phosphate, amino acid and tryptophan in legumes can influence the growth rate of *E. coli*. Specifically, glucose only influences the growth rate of *E. coli* during at concentration. For example, when glucose content is above 5 μM, the growth rate of *E. coli* increases slightly from 0.75 to 0.8 hour$^{-1}$. Between 0 and 5 μM, the growth rate of *E. coli* increased significantly from 0 to 0.75 hour$^{-1}$. The concentration of glucose, phosphate, amino acid and tryptophan in legumes have similar influence. If legumes have high phosphate content, the growth rate of *E. coli* is not influenced by the amino acid content. On the other hand, when the content of amino acids is below 1 μM, phosphate has a dramatic impact on the growth of *E. coli* (Shehata and Marr 1971).

Stantiall and collaborator (2018) and Serventi and collaborator (2018) observed that in legume cooking water, the main nutrients are soluble carbohydrates and insoluble carbohydrates. One fourth of Aquafaba dry matter is soluble carbohydrates (Stantiall et al. 2018). Therefore, the high sugar content of legume cooking water might have counteracted any inhibitory effect on *E. coli.*

## 7.4 Conclusions

The cooking water of legumes exhibited several functionalities. The focus of the current research was on texturizers, but a recent study has shown exciting potential as prebiotic ingredient. All legumes tested produced liquid ingredients of slightly acidic pH (around 6.0), low density and low to intermediate viscosity based on the fibre content. Despite the boiling process, the vast majority of the protein fraction of dry matter was soluble, contributing to foaming and emulsifying properties. Foaming abilities were moderate in comparison to a common foaming agent like egg white. Emulsifying properties were excellent, especially for chickpeas and lentils. Soluble protein, polysaccharides (mainly insoluble fibre like hemicellulose), saponins and surfactant phenolics likely played key roles in the development of highly stable emulsions. Gelling abilities were the result of moderate water absorption (chickpeas, peas and soy) and high oil absorption for all legumes. Fascinatingly, the cooking water of three pulses (beans, chickpeas and peas) resulted to be excellent prebiotics. They enhanced the growth of probiotic *lactobacilli* to a higher extent than a nutrient broth. Simultaneously, no antibacterial effects were observed for pathogens. In closing, legume cooking water have broad potential as texturizers: foaming agents, emulsifiers and gelling agents. Their applications as additive could significantly reduce the cost of food production and avoid the dependence on scarcely available ingredients. Furthermore, legume cooking water are promising prebiotic ingredients which could be used for the development of functional foods such as probiotic bars, beverages or fermented foods.

**Acknowledgments** Authors would like to thank Anirudh Sounderrajan for analysing the protein solubility of legume cooking water, as well as Yaying Luo for performing literature review on the antimicrobial properties of legumes. The courses FOOD 398, FOOD 399 and FOOD 699 of Lincoln University (New Zealand) provided the funding necessary for this research.

## References

Aguilar, N., Albanell, E., Miñarro, B., & Capellas, M. (2015). Chickpea and tiger nut flours as alternatives to emulsifier and shortening in gluten-free bread. *LWT-Food science and Technology, 62*(1), 225–232.

Aravantinos-Zafiris, G., Oreopoulou, V., Tzia, C., & Thomopoulos, C. D. (1994). Fibre fraction from orange peel residues after pectin extraction. *LWT-Food Science and Technology, 27*(5), 468–471.

Araya-Cloutier, C., Vincken, J. P., van Ederen, R., den Besten, H. M., & Gruppen, H. (2018). Rapid membrane permeabilization of *Listeria monocytogenes* and *Escherichia coli* induced by antibacterial prenylated phenolic compounds from legumes. *Food Chemistry, 240*, 147–155.

Aydemir, L. Y., & Yemenicioğlu, A. (2013). Potential of Turkish Kabuli type chickpea and green and red lentil cultivars as source of soy and animal origin functional protein alternatives. *LWT-Food Science and Technology, 50*(2), 686–694.

Bird, L. G., Pilkington, C. L., Saputra, A., & Serventi, L. (2017). Products of chickpea processing as texture improvers in gluten-free bread. *Food Science and Technology International, 23*(8), 690–698.

Bordenave, N., Hamaker, B. R., & Ferruzzi, M. G. (2014). Nature and consequences of non-covalent interactions between flavonoids and macronutrients in foods. *Food & Function, 5*(1), 18–34.

Böttcher, S., & Drusch, S. (2017). Saponins—Self-assembly and behavior at aqueous interfaces. *Advances in Colloid and Interface Science, 243*, 105–113.

Boye, J., Zare, F., & Pletch, A. (2010). Pulse proteins: Processing, characterization, functional properties and applications in food and feed. *Food Research International, 43*(2), 414–431.

Buhl, T. F., Christensen, C. H., & Hammershøj, M. (2019). Aquafaba as an egg white substitute in food foams and emulsions: Protein composition and functional behavior. *Food Hydrocolloids, 96*, 354–364.

Campbell, L., Raikos, V., & Euston, S. R. (2003). Modification of functional properties of egg-white proteins. *Food/Nahrung, 47*(6), 369–376.

Chen, M., Lu, J., Liu, F., Nsor-Atindana, J., Xu, F., Goff, H. D., Ma, J., & Zhong, F. (2019). Study on the emulsifying stability and interfacial adsorption of pea proteins. *Food Hydrocolloids, 88*, 247–255.

Chibbar, R. N., Ambigaipalan, P., & Hoover, R. (2010). Molecular diversity in pulse seed starch and complex carbohydrates and its role in human nutrition and health. *Cereal Chemistry, 87*(4), 342–352.

Chua, J. Y., & Liu, S. Q. (2019). Soy whey: More than just wastewater from tofu and soy protein isolate industry. *Trends in Food Science & Technology, 91*, 24–32.

Chua, J. Y., Lu, Y., & Liu, S. Q. (2018). Evaluation of five commercial non-*Saccharomyces* yeasts in fermentation of soy (tofu) whey into an alcoholic beverage. *Food Microbiology, 76*, 533–542.

Chung, C., Sher, A., Rousset, P., Decker, E. A., & McClements, D. J. (2017). Formulation of food emulsions using natural emulsifiers: Utilization of quillaja saponin and soy lecithin to fabricate liquid coffee whiteners. *Journal of Food Engineering, 209*, 1–11.

Damian, J. J., Huo, S., & Serventi, L. (2018). Phytochemical content and emulsifying ability of pulses cooking water. *European Food Research and Technology, 244*(9), 1647–1655.

de Fátima Viana, S., Guimarães, V. M., José, I. C., de Oliveira, M. G. D. A., Costa, N. M. B., de Barros, E. G., et al. (2005). Hydrolysis of oligosaccharides in soybean flour by soybean α-galactosidase. *Food Chemistry, 93*(4), 665–670.

Du, S. K., Jiang, H., Yu, X., & Jane, J. L. (2014). Physicochemical and functional properties of whole legume flour. *LWT-Food Science and Technology, 55*(1), 308–313.

Ettoumi, Y. L., Chibane, M., & Romero, A. (2016). Emulsifying properties of legume proteins at acidic conditions: Effect of protein concentration and ionic strength. *LWT-Food Science and Technology, 66*, 260–266.

European Commission. Horizon 2020. URL: https://ec.europa.eu/programmes/horizon2020/en. Accessed on 03 Oct 2019.

Felix, M., Cermeño, M., Romero, A., & FitzGerald, R. J. (2019). Characterisation of the bioactive properties and microstructure of chickpea protein-based oil in water emulsions. *Food Research International, 121*, 577–585.

Félix, M., Romero, A., Carrera-Sanchez, C., & Guerrero, A. (2019). A comprehensive approach from interfacial to bulk properties of legume protein-stabilized emulsions. *Fluids, 4*(2), 65.

Food Navigator USA. Most consumers want and will pay more for 'sustainable' options, but struggle to easily find them. 24-Jun-2019. By Elizabeth Crawford. URL: https://www.foodnavigator-usa.com/Article/2019/06/24/Most-consumers-want-and-will-pay-more-for-sustainable-options-but-struggle-to-easily-find-them. Accessed on 03 Oct 2019.

Foschia, M., Horstmann, S. W., Arendt, E. K., & Zannini, E. (2017). Legumes as functional ingredients in gluten-free bakery and pasta products. *Annual Review of Food Science and Technology, 8*, 75–96.

Garcia-Vaquero, M., Lopez-Alonso, M., & Hayes, M. (2017). Assessment of the functional properties of protein extracted from the brown seaweed Himanthalia elongata (Linnaeus) SF Gray. *Food Research International, 99*, 971–978.

Goto, T., Teraminami, A., Lee, J. Y., Ohyama, K., Funakoshi, K., Kim, Y. I., Hirai, S., Uemura, T., Yu, R., Takahashi, N., & Kawada, T. (2012). Tiliroside, a glycosidic flavonoid, ameliorates obesity-induced metabolic disorders via activation of adiponectin signaling followed by enhancement of fatty acid oxidation in liver and skeletal muscle in obese–diabetic mice. *The Journal of Nutritional Biochemistry, 23*(7), 768–776.

Güçlü-Üstündağ, Ö., & Mazza, G. (2007). Saponins: Properties, applications and processing. *Critical Reviews in Food Science and Nutrition, 47*(3), 231–258.

Gullón, B., Lú-Chau, T. A., Moreira, M. T., Lema, J. M., & Eibes, G. (2017). Rutin: A review on extraction, identification and purification methods, biological activities and approaches to enhance its bioavailability. *Trends in Food Science & Technology, 67*, 220–235.

Hassan, S. M., Byrd, J. A., Cartwright, A. L., & Bailey, C. A. (2010). Hemolytic and antimicrobial activities differ among saponin-rich extracts from guar, quillaja, yucca, and soybean. *Applied Biochemistry and Biotechnology, 162*(4), 1008–1017.

Jarpa-Parra, M. (2018). Lentil protein: A review of functional properties and food application. An overview of lentil protein functionality. *International Journal of Food Science & Technology, 53*(4), 892–903.

Jiang, J., Chen, J., & Xiong, Y. L. (2009). Structural and emulsifying properties of soy protein isolate subjected to acid and alkaline pH-shifting processes. *Journal of Agricultural and Food Chemistry, 57*(16), 7576–7583.

Kim, N. H., & Rhee, M. S. (2016). Phytic acid and sodium chloride show marked synergistic bactericidal effects against nonadapted and acid-adapted *Escherichia coli* O157: H7 strains. *Applied and Environmental Microbiology, 82*(4), 1040–1049.

Kutoš, T., Golob, T., Kač, M., & Plestenjak, A. (2003). Dietary fibre content of dry and processed beans. *Food Chemistry, 80*(2), 231–235.

Ladjal-Ettoumi, Y., Boudries, H., Chibane, M., & Romero, A. (2016). Pea, chickpea and lentil protein isolates: Physicochemical characterization and emulsifying properties. *Food Biophysics, 11*(1), 43–51.

Lafarga, T., Villaró, S., Bobo, G., & Aguiló-Aguayo, I. (2019). Optimisation of the pH and boiling conditions needed to obtain improved foaming and emulsifying properties of chickpea aquafaba using a response surface methodology. *International Journal of Gastronomy and Food Science, 18*, 100177.

Lian, H., Luo, K., Gong, Y., Zhang, S., & Serventi, L. (2019). Okara flours from chickpea and soy are thickeners: Increased dough viscosity and moisture content in gluten-free bread. *International Journal of Food Science & Technology, 55*(2), 805–812.

Marefati, A., Matos, M., Wiege, B., Haase, N. U., & Rayner, M. (2018). Pickering emulsifiers based on hydrophobically modified small granular starches Part II–Effects of modification on emulsifying capacity. *Carbohydrate Polymers, 201*, 416–424.

Martens, L. G., Nilsen, M. M., & Provan, F. (2017). Pea hull fibre: Novel and sustainable fibre with important health and functional properties. *EC Nutrition, 10*, 139–148.

Mateos-Aparicio, I., Pérez-López, E., & Rupérez, P. (2019). Valorisation approach for the soybean by-product okara using high hydrostatic pressure. *Current Nutrition & Food Science, 15*(6), 548–550.

Matsumiya, K., & Murray, B. S. (2016). Soybean protein isolate gel particles as foaming and emulsifying agents. *Food Hydrocolloids, 60*, 206–215.

Meurer, M. C., de Souza, D., & Marczak, L. D. F. (2019). Effects of ultrasound on technological properties of chickpea cooking water (aquafaba). *Journal of Food Engineering, 265*, 109688.

Mustafa, R., He, Y., Shim, Y. Y., & Reaney, M. J. (2018). Aquafaba, wastewater from chickpea canning, functions as an egg replacer in sponge cake. *International Journal of Food Science & Technology, 53*(10), 2247–2255.

Nikaido, H. (2003). Molecular basis of bacterial outer membrane permeability revisited. *Microbiology and Molecular Biology Review, 4*, 593–656.

Nilufer-Erdil, D., Serventi, L., Boyacioglu, D., & Vodovotz, Y. (2012). Effect of soy milk powder addition on staling of soy bread. *Food Chemistry, 131*(4), 1132–1139.

Olorunsola, E. O., Akpabio, E. I., Adedokun, M. O., & Ajibola, D. O. (2018). Emulsifying properties of hemicelluloses. In *Science and technology behind nanoemulsions* (p. 29), IntechOpen, London, UK.

Ostermann-Porcel, M. V., Rinaldoni, A. N., Rodriguez-Furlán, L. T., & Campderrós, M. E. (2017). Quality assessment of dried okara as a source of production of gluten-free flour. *Journal of the Science of Food and Agriculture, 97*(9), 2934–2941.

Ozturk, B., & McClements, D. J. (2016). Progress in natural emulsifiers for utilization in food emulsions. *Current Opinion in Food Science, 7*, 1–6.

Pal, R. S., Bhartiya, A., Yadav, P., Kant, L., Mishra, K. K., Aditya, J. P., & Pattanayak, A. (2017). Effect of dehulling, germination and cooking on nutrients, anti-nutrients, fatty acid composition and antioxidant properties in lentil (Lens culinaris). *Journal of Food Science and Technology, 54*(4), 909–920.

Pearce, K. N., & Kinsella, J. E. (1978). Emulsifying properties of proteins: Evaluation of a turbidimetric technique. *Journal of Agricultural and Food Chemistry, 26*(3), 716–723.

Pina-Pérez, M. C., & Pérez, M. F. (2018). Antimicrobial potential of legume extracts against foodborne pathogens: A review. *Trends in Food Science & Technology, 72*, 114–124.

Rodríguez, R., Jimenez, A., Fernández-Bolaños, J., Guillén, R., & Heredia, A. (2006). Dietary fibre from vegetable products as source of functional ingredients. *Trends in Food Science & Technology, 17*(1), 3–15.

Sanjeewa, W. T., Wanasundara, J. P., Pietrasik, Z., & Shand, P. J. (2010). Characterization of chickpea (Cicer arietinum L.) flours and application in low-fat pork bologna as a model system. *Food Research International, 43*(2), 617–626.

Serventi, L., Wang, S., Zhu, J., Liu, S., & Fei, F. (2018). Cooking water of yellow soybeans as emulsifier in gluten-free crackers. *European Food Research and Technology, 244*(12), 2141–2148.

Shehata, T. E., & Marr, A. G. (1971). Effect of nutrient concentration on the growth of Escherichia coli. *Journal of Bacteriology, 107*(1), 210–216.

Shi, L., Arntfield, S. D., & Nickerson, M. (2018). Changes in levels of phytic acid, lectins and oxalates during soaking and cooking of Canadian pulses. *Food Research International, 107*, 660–668.

Singhal, A., Karaca, A. C., Tyler, R., & Nickerson, M. (2016). *Pulse proteins: From processing to structure-function relationships* (p. 55). Grain Legumes, London, UK.

Stantiall, S. E., Dale, K. J., Calizo, F. S., & Serventi, L. (2018). Application of pulses cooking water as functional ingredients: The foaming and gelling abilities. *European Food Research and Technology, 244*(1), 97–104.

Subuola, F., Widodo, Y., & Kehinde, T. (2012). Processing and utilization of legumes in the tropics. In *Trends in vital food and control engineering* (pp. 71–85). Rijeka, IntechOpen.

Sui, X., Bi, S., Qi, B., Wang, Z., Zhang, M., Li, Y., & Jiang, L. (2017). Impact of ultrasonic treatment on an emulsion system stabilized with soybean protein isolate and lecithin: Its emulsifying property and emulsion stability. *Food Hydrocolloids, 63*, 727–734.

Tabtabaei, S., Konakbayeva, D., Rajabzadeh, A. R., & Legge, R. L. (2019). Functional properties of navy bean (Phaseolus vulgaris) protein concentrates obtained by pneumatic tribo-electrostatic separation. *Food Chemistry, 283*, 101–110.

Toews, R., & Wang, N. (2013). Physicochemical and functional properties of protein concentrates from pulses. *Food Research International, 52*(2), 445–451.

Tosh, S. M., & Yada, S. (2010). Dietary fibres in pulse seeds and fractions: Characterization, functional attributes, and applications. *Food Research International, 43*(2), 450–460.

Tyler, R., Wang, N., & Han, J. (2017). Composition, nutritional value, functionality, processing, and novel food uses of pulses and pulse ingredients. *Cereal Chemistry, 94*(1), 1–1.

Vong, W. C., & Liu, S. Q. (2019). The effects of carbohydrase, probiotic *Lactobacillus paracasei* and yeast *Lindnera saturnus* on the composition of a novel okara (soybean residue) functional beverage. *LWT, 100*, 196–204.

Xiong, T., Ye, X., Su, Y., Chen, X., Sun, H., Li, B., & Chen, Y. (2018). Identification and quantification of proteins at adsorption layer of emulsion stabilized by pea protein isolates. *Colloids and Surfaces B: Biointerfaces, 171*, 1–9.

Xu, M., Jin, Z., Ohm, J. B., Schwarz, P., Rao, J., & Chen, B. (2018). Improvement of the antioxidative activity of soluble phenolic compounds in chickpea by germination. *Journal of Agricultural and Food Chemistry, 66*(24), 6179–6187.

Zhang, B., Peng, H., Deng, Z., & Tsao, R. (2018). Phytochemicals of lentil (*Lens culinaris*) and their antioxidant and anti-inflammatory effects. *Journal of Food Bioactives, 1*, 93–103.

Zhou, Q. I., Zhao, Y. U., Dang, H., Tang, Y., & Zhang, B. (2019). Antibacterial effects of phytic acid against foodborne pathogens and investigation of its mode of action. *Journal of Food Protection, 82*(5), 826–833.

# Chapter 8
# Cooking Water Applications

**Luca Serventi, Yiding Yang, and Yaqi Bian**

## 8.1 Introduction

Legume cooking water is defined in this book chapter as the wastewater resulting from boiling and canning of legumes. The common term "Aquafaba" typically refers to the sole canning water of chickpeas (Mustafa et al. 2018). The original idea of using Aquafaba as food ingredient was the result of a culinary discovery attributed to a musician named Joël Roessel, dated December 2014. Brine from canned chickpeas was incorporated in a Swiss meringue recipe in replacement of egg white to produce a vegan alternative (Révolution végétale 2014). Subsequently, Aquafaba gained worldwide popularity and several recipes have been proposed. Vegan and non-vegan groups posted their ideas on dedicated websites. Nonetheless, no scientific information was available on the applications of legume cooking water in food product development until 2017.

Research in this field consists of seven published studies. Four of these peer-reviewed papers discussed applications in the confectionery industry: raw products such as French meringues and cream mousse (Damian et al. 2018; Meurer et al. 2019) and baked products called Swiss meringues (Lafarga et al. 2019; Stantiall et al. 2018). Three papers discussed applications of legume cooking water in bakery products: savoury (gluten-free bread and crackers) (Bird et al. 2017; Serventi et al. 2018) and sweet (sponge cake) (Mustafa et al. 2018). One study was recently published on Aquafaba use in mayonnaise (Lafarga et al. 2019).

Confectionery products require volume, soft structure, sweetness and appealing look which could be provided by legume cooking water due to the emulsifying, foaming and gelling abilities of Aquafaba carbohydrates, proteins and phytochemicals (Damian et al. 2018; Stantiall et al. 2018). Similarly, gluten-free bread and

L. Serventi (✉) · Y. Yang · Y. Bian
Department of Wine, Food and Molecular Biosciences, Faculty of Agriculture and Life Sciences, Lincoln University, Lincoln, Christchurch, New Zealand
e-mail: Luca.Serventi@lincoln.ac.nz



L. Serventi, *Upcycling Legume Water: from wastewater to food ingredients*,
https://doi.org/10.1007/978-3-030-42468-8_8

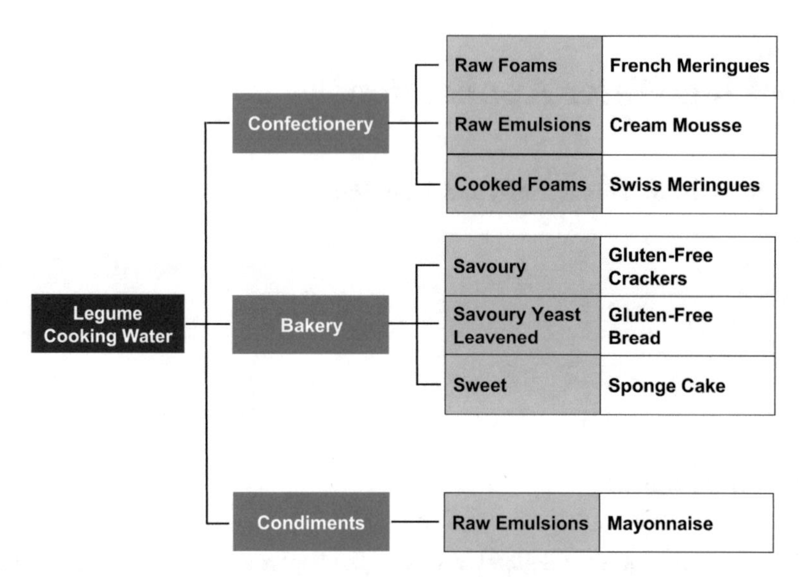


**Fig. 8.1** Applications of legume cooking water in food products (Bird et al. 2017; Damian et al. 2018; Lafarga et al. 2019; Meurer et al. 2019; Mustafa et al. 2018; Serventi et al. 2018; Stantiall et al. 2018)

sponge cake require volume, emulsifying ability (Sahi and Alava 2003) and hydrocolloid properties (Arslan et al. 2018; Nammakuna et al. 2016; Roman et al. 2019) which could be provided by Aquafaba fibre and protein (Serventi et al. 2018; Stantiall et al. 2018). Nonetheless, challenges may arise when replacing eggs with legumes: colour, aroma and flavour may change and prove to be unappealing. Despite similar quality features, the products listed above are manufactured through different processes. Raw confectionery require raw material with good aroma profile and sweet flavour; this could be a problem considering that most legumes exhibit "beany" features of bitter, earthy aroma and flavour (Roland et al. 2017). Batters for confectionery and cakes, as well as bread dough, must provide a stable structure during baking, without collapses. The main challenge with gluten-free crackers is the lack of structure, resulting in low puffiness and high chewiness (Nammakuna et al. 2016). The addition of fibre and protein isolates from soy and peas improved the texture of gluten-free crackers (Nammakuna et al. 2016) thus suggesting potential applications for Aquafaba. Bread inherently presents an added challenge due to presence of yeast, which is sensitive to the presence of salt, sugar and changes in pH (Romano and Capece 2010; Sahlström et al. 2004).

The following sections evaluate challenges and opportunities of using legume cooking water (from beans, chickpeas, lentils, peas and soy) in confectionery, bakery and spread applications (Fig. 8.1). Confectionery applications are presented in the following order: raw foams (French meringues), raw emulsions (cream mousse) and cooked foams (Swiss meringues). Bakery applications are presented as savoury (crackers), savoury yeast-leavened (bread) and sweet (sponge cake). In addition, an experimental section on mayonnaise discusses the result of a new study on Aquafaba applications in condiments, implemented with the results from a recently published paper (Lafarga et al. 2019).

## 8.2 Confectionery

### *8.2.1 Raw Foams (French Meringues)*

One of the simplest confectionery products is French meringues. A recent study evaluated the application of legume cooking water in uncooked, French style meringues (Meurer et al. 2019). Specifically, chickpeas were soaked as by Stantiall and collaborators (2018) and then heated by pressure cooker for 20 minutes instead of boiling. French meringues were manufactured by whipping the foaming agent (Aquafaba or egg white) first alone, then in combination with powdered sugar, in a 50:40 ratio (foaming agent:sugar) (Meurer et al. 2019). Instrumental evaluation of French meringues made with Aquafaba revealed significantly darker colour (L∗ 57 vs. 87). No explanations were provided for such change but it can be speculated that it was the results of pigments, mainly phenolic compounds, leached form chickpeas to their cooking water (Damian et al. 2018). Aquafaba French meringues were softer than the control (0.22 vs. 0.32 N). This was the result of a less cohesive (1.95 vs. 3.48 N.s) and less adhesive (0.10 vs. 0.39 N.s) structure (Meurer et al. 2019). Egg white contains heat coagulative proteins that confer extensibility to bakery products such as cake (Johnson and Zabik 1981). Aquafaba likely did not exhibit heat coagulative properties and this can be explained differences in protein structure between eggs and pulses (beans, chickpeas, lentils and peas). Egg white mainly consists of ovalbumin, abundant in free sulfhydryl groups (SH) that form intermolecular disulphide bonds (SS) upon heating, thus forming gels (Deleu et al. 2016). On the contrary, albumin represents only 10–20% of pulse proteins, with about 70% of globulins and 10% of other fractions (Singhal et al. 2016) thus resulting in lesser gelling. Furthermore, the protein leached into the pulse cooking water were likely of high solubility, thus not contributing to gelling. This hypothesis was confirmed by the same researchers, measuring solubility of 863 μg/ml for Aquafaba (Meurer et al. 2019) equal to 86.3% of the total protein. Interestingly, ultrasound treatment of Aquafaba significantly improved the quality of the final products, resulting in French meringues of similar colour and texture to the control: L∗ 84, hardness 0.26 N, consistency 5.67 N.s and adhesiveness 0.30 N.s (Meurer et al. 2019). Protein solubility was not affected by ultrasound treatment, but foaming and emulsifying abilities were. It is plausible that ultrasounds modified the protein structure of Aquafaba, increasing hydrophobic interactions without changes to solubility.

### *8.2.2 Raw Emulsions (Mousse)*

Confectionery products often consist of emulsions with sugar and different types of creams. One of the simplest confectionery emulsions is cream mousse, with the most popular type being chocolate mousse, an aerated dessert with stabilized foamy structure. Mousse manufacturing requires formation and stabilization of foam, as well as ingredients with emulsifying and stabilizing properties (Aragon-Alegro et al. 2007). The main ingredients of cream mousse are dairy cream, chocolate, sugar and either milk or egg white. Considering the overwhelming sensory characteristics of chocolate, Damian and collaborators (2018) tested legume cooking water in a plain cream mousse recipe, based only on foaming agent (egg white or legume cooking water) sugar and dairy cream. Authors used the cooking water of chickpeas and split soybeans as a foaming agent for mousse, comparing their emulsifying and foaming abilities to egg white. The two legumes tested yielded different results. All mousse recipes rated very high for overall acceptability, ranging from 6.5 and 6.9 of peas and chickpeas (like moderately) to 7.5 of the egg white control (like very much) (Fig. 8.2). Interestingly, appearance was not affected by the use of cooking water, with no changes to neither colour nor glossiness (Fig. 8.2). Chickpea and pea cooking water contained moderate levels of phenolics (~0.6 mg/g) but the high percentage of sugar in the recipe (66%) prevailed in appearance (Damian et al. 2018). Surprisingly, aroma was not affected by the change in foaming agent, being stable at ratings of ~6.5 (like moderately) (Fig. 8.2). The cooking water alone presented a moderate smell of cooked legume, which could be described as "nutty" (Miller et al. 2013) and possibly limited sensory acceptability. Nonetheless, the nutty aroma disappeared in the final products. It has been demonstrated that at high levels of aeration (80% v/v), sensory quality of foams is determined by the characteristics of its air bubbles, rather than by its matrix (Minor et al. 2009). The cooking water used by Damian ad collaborators possessed foaming ability of 58% and 95% (chickpeas and peas, respectively) (Stantiall et al. 2018) which can significantly

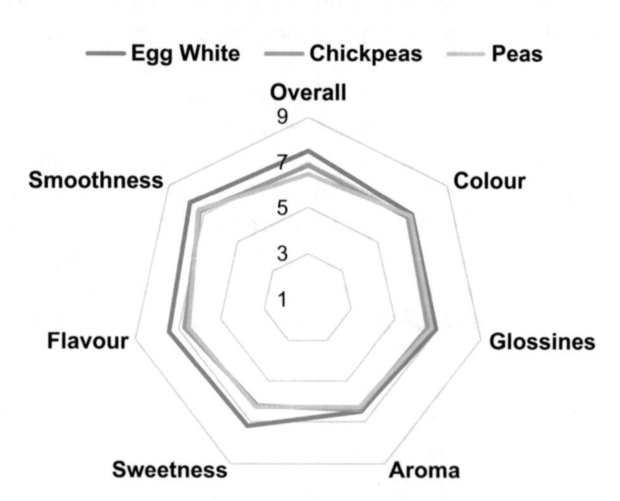


**Fig. 8.2** Sensory profile of cream mousse recipes made with either egg white, cooking water of chickpeas or cooking water of split yellow peas. (Data from Damian and collaborators, 2018)

increase the aeration level of the mousse. Furthermore, the compounds responsible for "beany", unpleasant aroma in legumes (1-octen-3-one, hexanal, 3-isopropyl-2-methoxypyrazine, 3-sec-butyl-2-methoxypyrazine, 3-isobutyl-2-methoxypyrazine and others) are volatile (Bott and Chambers IV 2006; Roland et al. 2017) thus easily prone to vaporise upon the long, intense whipping stage required for mousse manufacturing.

On the contrary, sweetness and flavour changed across the formulations tested. Sweetness decreased from 7.2 (like moderately) of control to about 6 (like slightly) of the legume mousse (Fig. 8.2). Sweetness decrease was not significant but it depicted a clear effect. The cooking water of chickpeas and peas contained high levels of saponins: 12 and 9.8 mg/g (Damian et al. 2018). Saponins are known to confer bitter flavour to food (Heng et al. 2006) and the bittering effect is particularly strong for soyasaponin I, the most common type found in legumes like chickpeas and peas (Singh et al. 2017). Therefore, the high saponin content may represent a challenge for application in sweet food products. Similarly, a reduction in flavour ratings was recorded, significantly for peas: 6.5 vs. 7.4 (Fig. 8.2). Pea saponins likely reduced flavour quality by means of bitterness, but other factors might have played a role: phenolic compounds (bitter and metallic flavour) and peptides/amino acids (metallic flavour) (Roland et al. 2017). Chickpea cooking water had lesser impact, probably due to a lower presence of unpleasant flavourings in comparison to peas. Uncommon isoflavones formononetin and biochanin A, have been associated to bitter flavour, and were detected in chickpeas at low levels (Roland et al. 2017).

Finally, no significant changes were observed for smoothness: ~7.3 vs. 7.8 (Fig. 8.2). Gelling abilities of the cooking water form chickpeas and split yellow peas were comparable to that of egg white, while foaming abilities were lower (58% and 93% vs. 400%) (Stantiall et al. 2018) consequently affecting smoothness of the mousse, but below perception levels. The cooking water of chickpeas and peas were successfully used to develop cream mousse, with minor concerns for peas due to the presence of bitter bioactives.

### *8.2.3 Cooked Foams (Swiss Meringues)*

Confectionery products can be baked. Therefore, meringues were chosen as matrix to investigate the feasibility of legume cooking water as foaming and gelling agent in sweet foods subjects to baking. This experimental setup matches the original idea of recycling canning water (or brine) from chickpeas as egg replacer to produce vegan meringues (Vegan Society of Aotearoa New Zealand 2019). Two studies tested the meringue application of legume cooking water: the first one evaluating the cooking water of haricot beans, chickpeas, green lentils, and split yellow peas (Stantiall et al. 2018) and the second one focusing on chickpeas (Lafarga et al. 2019).

Both studies conducted sensory analysis on 40 untrained subjects, revealing encouraging results. In the first study the overall acceptability was high for all

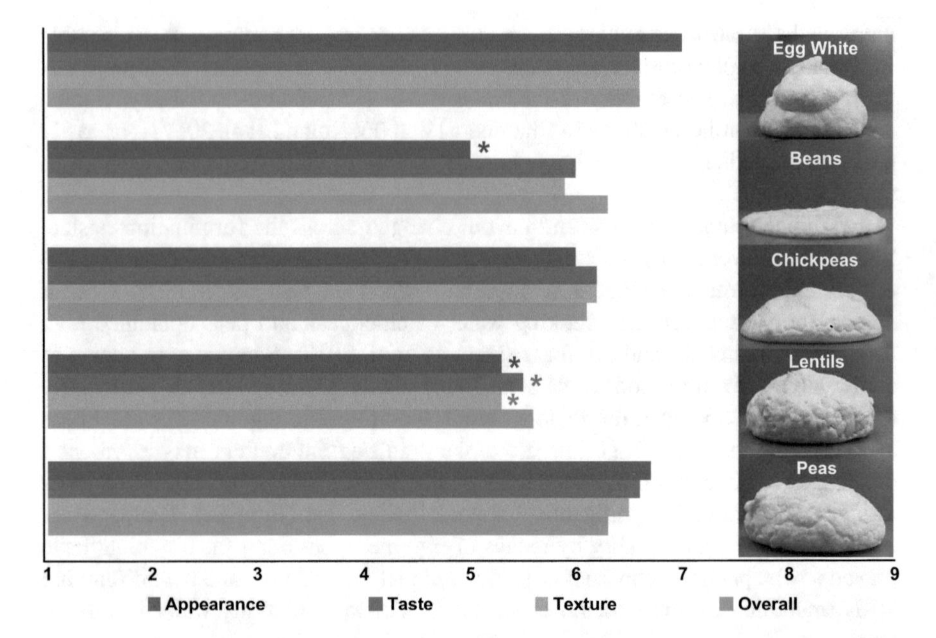


**Fig. 8.3** Descriptive sensory analysis of Swiss meringues made with egg white and cooking water of four legumes (data from Stantiall and others Stantiall et al. 2018). Asterisks refer to statistically significant difference within sensory attributes ($p < 0.05$). (Photo credits: Weihan Zhang (edited))

meringues, averaging around 6.6 on a 9-point hedonic scale (like moderately) (Stantiall et al. 2018) (Fig. 8.3). Chickpea and pea products rated similarly to the egg white control for all attributes: appearance, taste and texture (Fig. 8.3). On the contrary, challenges were depicted for beans and lentils. As shown by the picture in Fig. 8.3, the bean meringues was flat. Authors described development of volume during whipping of the mixture, but collapse in the oven at the start of the baking step. This negative result was attributed to poor gelling ability of the bean cooking water, with responsibility given to its low amount of solids, particularly insoluble fibre, in comparison to the other wastewater: 0.93 vs. ~2 g/100 g (Stantiall et al. 2018). Far more challenging was the situation of the lentil meringues, depicting significantly lower ratings for appearance, taste and texture (Fig. 8.3). Appearance of lentil meringues was evaluated 5.3 (neither like nor dislike) versus 7.0 (like moderately) of the control. Panellists commented on the colour as too dark. This colour change was ascribed to the phenolic compounds and tannins, of which lentils are an abundant source (Mirali et al. 2017). Bitterness was another recurring comment by panellists, causing the taste rating of lentil meringues to drop from 6.6 of the control to 5.5 (Fig. 8.2). Not only phenolics, but also saponins are highly abundant in lentils (del Hierro et al. 2018; Zhang et al. 2018) and in lentil cooking water (Damian et al. 2018), thus conferring bitter, astringent and earthy flavour, which do not match the traditional sweet flavour of meringues (Jesús et al. 2013). Interestingly, panellists did not like the texture of lentil meringues, assigning it a rating of 5.3 vs. 6.6 of control (Fig. 8.3). Excessive water absorption by fibre and/or protein was speculated to be the reason. A later study (Damian et al. 2018) disagreed with this hypothesis, showing that lentil cooking water absorbed less water and similar amount of oil than the other ingredients. Therefore, it is plausible that water distribution, rather than quantity, affected meringues stickiness.

Instrumental analysis supported the sensory profiles described above. Moisture content was lower for all legume meringues (~5 vs. 10 g/100 g of egg white) likely due to their lower content of dry matter in comparison to egg white: 3.3–5.1 vs. ~10 g/100 g (Stantiall et al. 2018; Uysal et al. 2019). Interestingly, most legume meringues were characterised by higher specific volume (~12 vs. 6.1 ml/g) despite similar height. Legume cooking water were shown to be less dense than egg white: 1.017–1.025 vs. 1.040 g/ml (Stantiall et al. 2018) likely explaining the development of larger, lighter structures. This hypothesis was confirmed by texture analysis. While meringue hardness was comparable to egg white for two ingredients (chickpeas and lentils), extensibility was drastically lower for all new products: ~2.0 vs. 14 mm (Stantiall et al. 2018). As discussed in Sect. 8.2.1 for French meringues, pulse proteins do not possess the same gelling ability of egg white due to a lower content of albumin thus resulting in lesser gelling and high protein solubility (Meurer et al. 2019) which is perceived as lower extensibility (Stantiall et al. 2018). It is hypothesized that the protein fraction of pulses leached in the cooking water was highly soluble and that it consisted of more globulin than albumin. Pulses cooking water are water solutions of low concentration (3–5% dry matter) (Stantiall et al. 2018) thus heating in diluted solvent did not alter the conformation of pulse proteins.

Finally, colour differences were observed for chickpeas and lentils, being greener and yellower than the other meringues: $a*$ −7 vs. ~0 and $b*$ 14 vs. ~5, respectively, attributed to chlorophylls and phenolics (Stantiall et al. 2018). Nonetheless, panellists were not affected by these differences. Considering the overall success, but the visual and texture limitations of chickpea cooking water, Lafarga and collaborators (2019) optimised its pH and boiling conditions. In terms of boiling, the ratio of raw chickpeas to water ranged from 1:1.5 to 1:50 (Lafarga et al. 2019) whereas the previous study used a 1:1.75 ratio (soaked chickpeas to water) (Stantiall et al. 2018). Once the cooking water was obtained, pH was adjusted with lemon juice, ranging from 3.5 to 6.5 (as opposed to the pH value of 6.3 observed by Stantiall and collaborators, 2018). Authors identified the optimal conditions for foaming and emulsifying properties in a cooking ratio of 1:1.50 chickpeas:water and cooking water pH of 3.50 (Lafarga et al. 2019). Meringues developed with the optimal ingredient led to superior visual quality (sensory rating 7.7 vs. 7.6 of egg white) and texture acceptance (8.0 vs. 7.5 of egg white) (Lafarga et al. 2019), while no changes occurred to flavour and overall acceptability. These results were attributed to the enhanced foaming and emulsifying properties of chickpea cooking water (Lafarga et al. 2019).

## 8.3 Bakery

### 8.3.1 *Savoury (Crackers)*

Savoury bakery products can be manufactured without yeast. For example, leavening of crackers is commonly achieved with inorganic salts which, when added to the dough, produce gases for textural development (Manley 2011). The absence of yeast removes the biological variable, allowing food technologists to focus on the physicochemical properties of the ingredients tested. One study applied the cooking water of yellow soybeans to the manufacturing of gluten-free crackers, replacing all tap water in the formula with it (Serventi et al. 2018). Changes to texture and moisture during a short 2-day storage at room temperature were observed and results were extremely interesting. While the control cracker expectedly doubled in hardness over time, the soy cracker softened by three folds (Table 8.1). Consequently, 2-days old crackers reported hardness values of 40 kg (control) and 8.2 kg (soy) (Table 8.1). Furthermore, the moisture content of the soy cracker doubled (from 12 to 25 g/100 g) in contrast with a steady value of 14 g/100 g of the control (Table 8.1). The RVA analysis revealed changes in pasting properties, with significantly higher breakdown viscosity in soy than control (24 vs. 21 RV), drastically lower setback viscosity (almost half, 32 vs. 59 RVU) and longer peak time (7.0 vs. 6.3 minutes) (Serventi et al. 2018). Lower setback is typically correlated to less starch recrystallization upon cooling and storage (Collar 2003). Longer gelatinisation time depicts enhanced water absorption by non-starch ingredients (Chaisawang and Suphantharika 2006). Consequently, the data collected suggested that the solids found in soy cooking water prevented amylopectin recrystallisation and enhanced water retention, promoting moisture absorption from the storage environment (plastic bag at room temperature). Soy cooking water showed a moderate water absorption capacity (1.54 g/g) (Serventi et al. 2018). Nonetheless, soy protein isolates were demonstrated to absorb more water than pulse proteins (chickpeas, faba beans, lentils and peas) (Gwiazda et al. 1979; Jarpa-Parra 2018; Withana-Gamage et al. 2011) thus it is plausible that the low amount of protein found in soy cooking water (0.68 g/100 g) (Serventi et al. 2018) slowly absorbed moisture over time, resulting in this outstanding moistening, softening effect on gluten-free crackers (Table 8.1).

**Table 8.1** Hardness and moisture content of gluten-free crackers made with tap water (Control) and the cooking water of yellow soybeans (Soy Cooking Water)

| \ | Control | Soy cooking water |
|---|---|---|
| Hardness (kg) | | |
| *Day 0* | $20^{b}$ | $26^{b}$ |
| *Day 2* | $40^{a}$ | $8.2^{c}$ |
| Moisture content (g/100 g) | | |
| *Day 0* | $17^{b}$ | $12^{c}$ |
| *Day 2* | $14^{bc}$ | $25^{a}$ |

Data from Serventi and others, 2018.
Different letters refer to statistically significant difference within instrumental parameter ($p < 0.05$)

Staling is a complex phenomenon that involves starch retrogradation (amylopectin recrystallisation), moisture loss and reduced protein plasticisation (Gray and Bemiller 2003). Based on these findings it can be speculated that the cooking water of yellow soybeans delayed staling of gluten-free crackers by means of reduced starch retrogradation and increased water absorption. Such hypothesis finds confirmation in a study on the effect of soy milk powder addition to soy bread. This study observed reduced moisture loss upon storage of bread formulated with insoluble fibres from soy (Nilufer-Erdil et al. 2012). Insoluble fibre was the main solid fraction of soy cooking water, representing about half of it (Serventi et al. 2018) and it was demonstrated to absorb high amounts of water (Chen et al. 2014). Soy protein isolates were shown to have less effect on water dynamics, with the exception of insoluble (denatured) protein which resulted in lower water mobility (more "unfreezable" water) (Nilufer-Erdil et al. 2012). Therefore, the lower setback value may be ascribed to the ratio of solids found in soy cooking water: 74% carbohydrates (44% insoluble and 30% soluble), 14% ash and 12% protein (Serventi et al. 2018). In comparison, soy milk powder contained drastically less carbohydrates (31%, of which 7% fibre), half the ash (6%), four times more protein (48%, mostly denatured) and more fat (12%) (Benesoy, USA). Overall, the beneficial antistaling effects of soy cooking water could be attributed to a high amount of insoluble fibre, combined to a low amount of soluble fibre and a fraction of proteins.

Interestingly, a study on rice-based gluten-free crackers tested hydrocolloids (carboxymethyl cellulose CMC, hydroxypropyl methylcellulose HPMC and xanthan gum) and legume protein (pea and soy), alone and in combination, as texture improvers (Nammakuna et al. 2016). Findings reported that neither hydrocolloids nor legume protein improved the texture of gluten-free crackers. Alternatively, combinations of hydrocolloids, pea and soy protein resulted in higher interaction with the rice base, increasing dough viscoelasticity, partially mimicking that of gluten-containing crackers. Specifically, the elastic modulus (G') increased, developing a rubbery dough which facilitated kneading and sheeting (Nammakuna et al. 2016). Authors however noticed undesirable flavours when soy and pea protein isolates were added. This challenge can be overcome with the use of soy cooking water due to the fact that boiling allowed removal of some of the undesired aromatic components of soybeans (Roland et al. 2017).

### *8.3.2 Savoury Yeast-Leavened (Bread)*

The use of yeast as leavening agent adds complexity to the science and technology of bakery products because baker's yeast (*Saccharomyces cerevisiae*) has specific substrate requirements (Romano and Capece 2010) that might limit the choice of suitable food ingredients. One study investigated the feasibility of using legume cooking water in breadmaking. Specifically, a gluten-free bread recipe based on rice flour and corn starch was modified by replacing tap water with chickpea cooking water. Instrumental analysis highlighted softer texture of the chickpea bread crumb:

3 vs. 4.3 kg (Bird et al. 2017). This was a noteworthy 30% reduction of hardness, which is a significant improvement for gluten-free bread, considering that this type of product is often too hard (Martínez and Gómez 2017; Stantiall and Serventi 2018). Legume flours have been proven successful in softening gluten-free bread (Melini et al. 2017) so this result was expected due to the similarity in composition between flour and cooking solids from chickpeas: 70–75% carbohydrates (50–55% insoluble and 20% soluble), 20–24% protein and 10% of other nutrients (Sreerama et al. 2012; Stantiall et al. 2018). In addition, the amount of chickpea cooking water used was 84 g for a 384 g dough, equal to 22% of the recipe. Considering its solid content of 5.13 g/100 g (Stantiall et al. 2018) the dose of solids added was 1.1% of the total recipe, equal to 2% on flour base. A 2% addition flour base is a common dosage for hydrocolloids and other bread improvers. Therefore, the solids of chickpea cooking water acted as additives, providing foaming, emulsifying and thickening abilities (Damian et al. 2018; Stantiall et al. 2018) that soften bread crumb. Figure 8.4 shows cross sections of the bread loves and micrographs from scanning electron microscopy of the inner surface of bread crumb pores. It is visible that the control gluten-free bread presents deep holes, which probably allowed the gas produced by yeast fermentation to escape during baking, resulting in hard texture (Fig. 8.4). On the contrary, the bread made with chickpea cooking water presented less and shallow holes, with a smooth surface (Fig. 8.4). The improved crumb

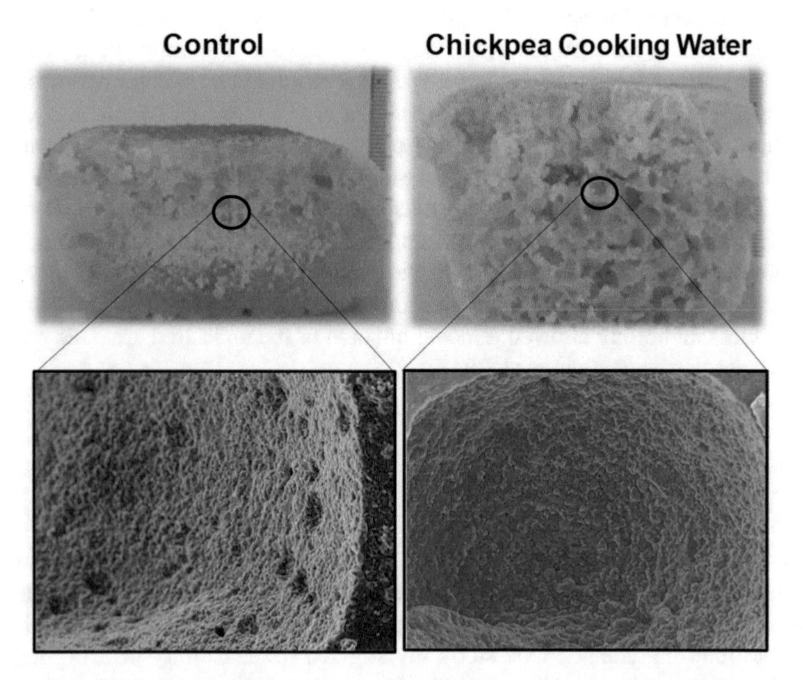


**Fig. 8.4** Top: pictures of gluten-free bread, control (left) and formulated with chickpea cooking water (right). Bottom: SEM micrographs of the internal surface of crumb pores, magnification 400×, for control (left) and chickpea cooking water (right). (Photos from Bird and collaborators (2017))

structure of the chickpea-containing bread could be the result of hydrocolloid properties of the fibres and protein found therein, resulting in more homogenous moisture distribution within the starch-protein network (Horstmann et al. 2018) thus enhanced gas retention leading to softer texture.

Water dynamics were affected by this ingredient. Pasting properties measured by a Rapid Visco Analyzer (RVA) changed, with significantly lower setback (51 vs. 63 RVU) and longer peak time (6.6 vs. 6.2 minutes) (Bird et al. 2017). As discussed in Sect. 8.3.1, lower setback is associated to lower starch recrystallization (Collar 2003) while longer peak time suggests stronger water entanglement in the starch-protein network (Chaisawang and Suphantharika 2006). These changes can delay staling and consequent bread hardening (Sun et al. 2019) thus storage studies are warranted.

### *8.3.3 Sweet (Sponge Cake)*

The increasing demand for vegan foods and flexitarian options (Gilligan 2019; Hood 2019) prompted researchers to find alternatives to egg white without sacrificing taste, appearance and texture. Scientists are gradually discovering the value of canned beans wastewater (Aquafaba). In a study by Mustafa and collaborators (2018) ten commercial brands of canned chickpeas were purchased at a grocery store. Cooked chickpeas were drained from the liquid and Aquafaba was collected to be used as foaming ingredient in sponge cake: challenges arose. First, the Aquafaba batter looked granular and irregular. Considering its diverse composition (Stantiall et al. 2018) it is possible that some components, perhaps insoluble fibre from the seed coat, did not mix homogeneously with the batter ingredients. Second, the cake loaf showed a minor collapse in the centre and no clear division took place between crumb and crust, indicating less structural stability (Mustafa et al. 2018). Two recent studies highlighted the inability of Aquafaba protein to denature and gel (Meurer et al. 2019; Shim et al. 2018) having high heat stability which could lead to weaker gel strength in cake. This limitation was overcome by ultrasound treatment of Aquafaba, which partially denatured it, thus enhancing hydrophobic interactions and gelling ability (Meurer et al. 2019). In addition, pH modifications were proposed as means of optimising the emulsifying ability of Aquafaba (Buhl et al. 2019). In terms of cake quality, crust colour was slightly darker ($L*$ 55 vs. 59) (Mustafa et al. 2018). Authors hypothesised that these colour changes were due to higher caramelisation degree (Maillard's reaction), attributed to higher polysaccharide content in Aquafaba than in egg white. On the contrary, crumb colour did not change; expectedly since Maillard's reaction only takes place in low moisture environment such as the crust (Ellis 1959). This result was in agreement with findings for bread crumb colour of bread made with chickpea cooking water (Bird et al. 2017). Finally, texture analysis revealed comparable cake hardness, but lower chewiness and springiness for the Aquafaba sample (Mustafa et al. 2018). As discussed above, Aquafaba protein exhibited lower gelling ability than egg white, likely developing a crumb that was less chewy and less elastic. Similar observations were reported for Swiss meringues made with chickpea cooking water: similar hardness than egg white, but less extensibility (Stantiall et al. 2018).

## 8.4 Condiments

### *8.4.1 Raw Emulsions (Mayonnaise)*

#### 8.4.1.1 Introduction

Lastly, Aquafaba was used to develop egg-free mayonnaise. Only one published study was found by Lafarga and collaborators (2019). The canning water of chickpeas was used to replace egg white in this food emulsion, based on sunflower oil. Colour and sensory analyses revealed darker colour for the Aquafaba product. Sensory analysis by 40 untrained panellists revealed lower acceptability index (80% vs. 92%). This result was mainly driven by appearance, whereas textural difference was not significant: 8.0 vs. 8.4, on a 9-point hedonic scale (Lafarga et al. 2019).

#### 8.4.1.2 Materials and Methods

Another unpublished study used Aquafaba to replace whole egg, to mimic common mayonnaise sold in supermarkets, which are based on vegetable oil and whole egg. A volume of 155 ml of Aquafaba or egg was mixed with 45 ml lemon juice, 15 ml vinegar (Pams, New Zealand), 15 ml brown rice syrup (Ceres Organics, New Zealand), 7.6 g salt (Pams, New Zealand) and 4.7 g mustard powder (Colman's, England) with a blender (Nutribullet Select, N9C-0907, USA) at speed 2 for 1 minute. Later, 400 ml of canola oil (Pams, New Zealand) were added into the mixture in three steps, then blended at speed 4 for 5 minutes. Colour was measured with a colorimeter (Minolta, CR-210) set on CIELAB scale. Density was determined with a hydrometer (Peter Stevenson LTD, Scotland). Viscosity of Aquafaba and egg mayonnaise was measured by Rapid Viscos analyser (RVA Super 4, Newport Scientific, New Zealand): 12.4 g of water and mayonnaise (50:50) were tested at 30 °C and viscosity recorded in triplicate after 1 minute.

#### 8.4.1.3 Results and Discussion

This experiment showed a colour difference between Aquafaba and egg mayonnaise (Table 8.2). The $L*$ of Aquafaba mayonnaise was 84, higher than that of egg (82), likely explained by the darker colour imparted by egg yok (Table 8.2). In addition, the $b*$ value of Aquafaba was lower than egg: 10 vs. 22, indicating less yellow colour (Table 8.2). Finally, the $a*$ value of Aquafaba was −0.68, lower than a near

**Table 8.2** Evaluation of mayonnaise products made by replacing egg ingredients with Aquafaba. Different letters refer to statistically significant difference within instrumental parameter ($p < 0.05$)

| | Aquafaba mayonnaise | Egg mayonnaise | References |
|---|---|---|---|
| Colour L∗ | 84–85 | 82–89 | Lafarga et al. (2019) |
| Colour a∗ | −3.4/−0.7 | −0.0/−3.2 | Lafarga et al. (2019) |
| Colour b∗ | 10–13 | 9.9–22 | Lafarga et al. (2019) |
| Colour ΔE | 4.95–12.5 | | Lafarga et al. (2019) |
| Sensory Overall Acceptance Score | 7.2 ± 0.7 | 8.3 ± 0.5 | Lafarga et al. (2019) |
| Density (g/ml) | 0.97 ± 0.06 | 0.92 ± 0.05 | *Bian & Serventi* |
| Viscosity (cP) | 13332 ± 341 | 14228 ± 803 | *Bian & Serventi* |

neutral value for the control (−0.01), thus indicating less yellow, as expected due to the darker yellow colour of egg yolk (Table 8.2). Density of the two mayonnaise products did not significantly vary: 0.97 vs. 0.92 g/ml (Table 8.2). Aquafaba contains complexed carbohydrates (Stantiall et al. 2018) which could increase the density of its emulsion. Finally, viscosity of diluted Aquafaba mayo was 13332 cp, lower than the 14228 cp of egg control, yet not significantly (Table 8.2). Eggs have better foaming volume and foaming ability than Aquafaba (Mustafa et al. 2018; Stantiall et al. 2018) which could incorporate with oil better to reduce the free liquid in mayonnaise. Overall, colour and textural results agreed with the published literature (Lafarga et al. 2019) showing small colour differences and loss of viscosity of Aquafaba mayonnaise. On a positive note, these challenges were shown to be small and likely to be overcome with further product development, such as use of yellow colouring and concentration of Aquafaba for higher viscosity.

## 8.5 Conclusions

The applications of legume cooking water in food started with the culinary discovery that Aquafaba (brine from canned chickpeas) can act as egg replacer to produce vegan meringues. Shortly after, the popularity of Aquafaba grew and inspired research on the cooking water of several legumes (haricot beans, chickpeas, green lentils, split yellow peas and yellow soybeans). Two areas of use were identified:

- Egg replacer (for flexitarians, vegans and those with egg allergy);
- Hydrocolloid (affordable and sustainable).

Legume cooking water was proven successful as egg replacer in French meringues, cream mousse and Swiss meringues. Particularly, chickpeas produced the best results in terms of structure and sensory profile. Furthermore, chickpea cooking water successfully replaced egg white in sponge cake and whole egg in mayonnaise. Minor challenges were identified, mainly dealing with inferior chewiness and colour differences. This problem was attributed to the heat stabile protein found in the ingredient, which thus develop weaker gels than egg protein under heat. Consequently, optimisation of the cooking conditions (seed to water ratio) has been proposed. In addition, physical treatment (ultrasound) and chemical treatment (lower pH) of the cooking water have been proposed to overcome this challenge, improving the emulsifying, foaming and gelling abilities. Colour can be addressed with colouring agents, while viscosity can be optimised by using powdered version of this liquid ingredient, thus allowing technologist to choose the desired concentration.

Chickpea cooking water also softened gluten-free bread by means of homogenous moisture distribution across the crumb. Furthermore, soy cooking water proved to be a strong antistaling agent in gluten-free crackers, promoting water absorption during storage and preventing starch retrogradation. These findings project legume cooking water as affordable and sustainable egg replacers and hydrocolloids for several food applications, promoting the concepts of circular economy and upcycling.

**Acknowledgments** Part of this research was funded by Lincoln University with the teaching budget allocated to the Bachelor course "FOOD 399 – Research Placement" and the taught Master course "FOOD 698 – Research Essay"

## References

Aragon-Alegro, L. C., Alegro, J. H. A., Cardarelli, H. R., Chiu, M. C., & Saad, S. M. I. (2007). Potentially probiotic and synbiotic chocolate mousse. *LWT-Food Science and Technology, 40*(4), 669–675.

Arslan, M., Rakha, A., Xiaobo, Z., & Mahmood, M. A. (2018). Complimenting gluten free bakery products with dietary fiber: Opportunities and constraints. *Trends in Food Science & Technology, 83*, 194–202.

Bird, L. G., Pilkington, C. L., Saputra, A., & Serventi, L. (2017). Products of chickpea processing as texture improvers in gluten-free bread. *Food Science and Technology International, 23*(8), 690–698.

Bott, L., & Chambers, E., IV. (2006). Sensory characteristics of combinations of chemicals potentially associated with beany aroma in foods. *Journal of Sensory Studies, 21*(3), 308–321.

Buhl, T. F., Christensen, C. H., & Hammershøj, M. (2019). Aquafaba as an egg white substitute in food foams and emulsions: Protein composition and functional behavior. *Food Hydrocolloids, 96*, 354–364.

Chaisawang, M., & Suphantharika, M. (2006). Pasting and rheological properties of native and anionic tapioca starches as modified by guar gum and xanthan gum. *Food Hydrocolloids, 20*(5), 641–649.

Chen, Y., Ye, R., Yin, L., & Zhang, N. (2014). Novel blasting extrusion processing improved the physicochemical properties of soluble dietary fiber from soybean residue and in vivo evaluation. *Journal of Food Engineering, 120*, 1–8.

Collar, C. (2003). Significance of viscosity profile of pasted and gelled formulated wheat doughs on bread staling. *European Food Research and Technology, 216*(6), 505–513.

Damian, J. J., Huo, S., & Serventi, L. (2018). Phytochemical content and emulsifying ability of pulses cooking water. *European Food Research and Technology, 244*(9), 1647–1655.

del Hierro, J. N., Herrera, T., García-Risco, M. R., Fornari, T., Reglero, G., & Martin, D. (2018). Ultrasound-assisted extraction and bioaccessibility of saponins from edible seeds: quinoa, lentil, fenugreek, soybean and lupin. *Food research international, 109*, 440–447.

Deleu, L. J., Wilderjans, E., Van Haesendonck, I., Brijs, K., & Delcour, J. A. (2016). Protein network formation during pound cake making: The role of egg white proteins and wheat flour gliadins. *Food Hydrocolloids, 61*, 409–414.

Ellis, G. P. (1959). The maillard reaction. In *Advances in carbohydrate chemistry* (Vol. 14, pp. 63–134). Academic Press, New York, NY, USA.

Gilligan, J. M. (2019). Modelling diet choices. *Nature Sustainability, 2*(8), 661–662.

Gray, J. A., & Bemiller, J. N. (2003). Bread staling: Molecular basis and control. *Comprehensive Reviews in Food Science and Food Safety, 2*(1), 1–21.

Gwiazda, S., Rutkowski, A., & Kocoń, J. (1979). Some functional properties of pea and soy bean protein preparations. *Food/Nahrung, 23*(7), 681–686.

Heng, L., Vincken, J. P., van Koningsveld, G., Legger, A., Gruppen, H., van Boekel, T., Roozen, J., & Voragen, F. (2006). Bitterness of saponins and their content in dry peas. *Journal of the Science of Food and Agriculture, 86*(8), 1225–1231.

Hood, S. (2019). Vegetarianism and vegan diets. In *Manual of dietetic practice* (p.129), John Wiley and Sons, Hoboken, NJ, USA.

Horstmann, S. W., Axel, C., & Arendt, E. K. (2018). Water absorption as a prediction tool for the application of hydrocolloids in potato starch-based bread. *Food Hydrocolloids, 81*, 129–138.

Jarpa-Parra, M. (2018). Lentil protein: a review of functional properties and food application. An overview of lentil protein functionality. *International Journal of Food Science & Technology, 53*(4), 892–903.

Jesús, M. N. D., Zanqui, A. B., Valderrama, P., Tanamati, A., Maruyama, S. A., Souza, N. E. D., & Matsushita, M. (2013). Sensory and physico-chemical characteristics of desserts prepared with egg products processed by freeze and spray drying. *Food Science and Technology, 33*(3), 549–554.

Johnson, T. M., & Zabik, M. E. (1981). Egg albumen proteins interactions in an angel food cake system. *Journal of Food Science, 46*(4), 1231–1236.

Lafarga, T., Villaró, S., Bobo, G., & Aguiló-Aguayo, I. (2019). Optimisation of the pH and boiling conditions needed to obtain improved foaming and emulsifying properties of chickpea aquafaba using a response surface methodology. *International Journal of Gastronomy and Food Science, 18*, 100177.

Manley, D. (Ed.). (2011). *Manley's technology of biscuits, crackers and cookies*. Elsevier, Sawston (England).

Martínez, M. M., & Gómez, M. (2017). Rheological and microstructural evolution of the most common gluten-free flours and starches during bread fermentation and baking. *Journal of Food Engineering, 197*, 78–86.

Melini, F., Melini, V., Luziatelli, F., & Ruzzi, M. (2017). Current and forward-looking approaches to technological and nutritional improvements of gluten-free bread with legume flours: A critical review. *Comprehensive Reviews in Food Science and Food Safety, 16*(5), 1101–1122.

Meurer, M. C., de Souza, D., & Marczak, L. D. F. (2019). Effects of ultrasound on technological properties of chickpea cooking water (aquafaba). *Journal of Food Engineering, 265*, 109688.

Miller, A. E., Chambers, E., IV, Jenkins, A., Lee, J., & Chambers, D. H. (2013). Defining and characterizing the "nutty" attribute across food categories. *Food Quality and Preference, 27*(1), 1–7.

Minor, M., Vingerhoeds, M. H., Zoet, F. D., De Wijk, R., & Van Aken, G. A. (2009). Preparation and sensory perception of fat-free foams–effect of matrix properties and level of aeration. *International Journal of Food Science & Technology, 44*(4), 735–747.

Mirali, M., Purves, R. W., & Vandenberg, A. (2017). Profiling the phenolic compounds of the four major seed coat types and their relation to color genes in lentil. *Journal of Natural Products, 80*(5), 1310–1317.

Mustafa, R., He, Y., Shim, Y. Y., & Reaney, M. J. (2018). Aquafaba, wastewater from chickpea canning, functions as an egg replacer in sponge cake. *International Journal of Food Science & Technology, 53*(10), 2247–2255.

Nammakuna, N., Barringer, S. A., & Ratanatriwong, P. (2016). The effects of protein isolates and hydrocolloids complexes on dough rheology, physicochemical properties and qualities of gluten-free crackers. *Food Science & Nutrition, 4*(2), 143–155.

Nilufer-Erdil, D., Serventi, L., Boyacioglu, D., & Vodovotz, Y. (2012). Effect of soy milk powder addition on staling of soy bread. *Food Chemistry, 131*(4), 1132–1139.

Révolution végétale. (2014). URL: http://www.revolutionvegetale.com/en/non-classe/la-mousse-vegetale/. Accessed on 30 Aug 2019.

Roland, W. S., Pouvreau, L., Curran, J., van de Velde, F., & de Kok, P. M. (2017). Flavor aspects of pulse ingredients. *Cereal Chemistry, 94*(1), 58–65.

Roman, L., Belorio, M., & Gomez, M. (2019). Gluten-free breads: The gap between research and commercial reality. *Comprehensive Reviews in Food Science and Food Safety, 18*(3), 690–702.

Romano, P., & Capece, A. (2010). Saccharomyces cerevisiae as Bakers' Yeast. In *Encyclopedia of biotechnology in agriculture and food* (pp. 1–4). CRC Press, New York, NY, USA.

Sahi, S. S., & Alava, J. M. (2003). Functionality of emulsifiers in sponge cake production. *Journal of the Science of Food and Agriculture, 83*(14), 1419–1429.

Sahlström, S., Park, W., & Shelton, D. R. (2004). Factors influencing yeast fermentation and the effect of LMW sugars and yeast fermentation on hearth bread quality. *Cereal Chemistry, 81*(3), 328–335.

Serventi, L., Wang, S., Zhu, J., Liu, S., & Fei, F. (2018). Cooking water of yellow soybeans as emulsifier in gluten-free crackers. *European Food Research and Technology, 244*(12), 2141–2148.

Shim, Y. Y., Mustafa, R., Shen, J., Ratanapariyanuch, K., & Reaney, M. J. (2018). Composition and properties of aquafaba: Water recovered from commercially canned chickpeas. *JoVE (Journal of Visualized Experiments), 132*, e56305.

Singh, B., Singh, J. P., Singh, N., & Kaur, A. (2017). Saponins in pulses and their health promoting activities: A review. *Food Chemistry, 233*, 540–549.

Singhal, A., Karaca, A. C., Tyler, R., & Nickerson, M. (2016). *Pulse proteins: From processing to structure-function relationships* (p. 55). Grain Legumes, London, England.

Sreerama, Y. N., Sashikala, V. B., Pratape, V. M., & Singh, V. (2012). Nutrients and antinutrients in cowpea and horse gram flours in comparison to chickpea flour: Evaluation of their flour functionality. *Food Chemistry, 131*(2), 462–468.

Stantiall, S. E., & Serventi, L. (2018). Nutritional and sensory challenges of gluten-free bakery products: A review. *International Journal of Food Sciences and Nutrition, 69*(4), 427–436.

Stantiall, S. E., Dale, K. J., Calizo, F. S., & Serventi, L. (2018). Application of pulses cooking water as functional ingredients: The foaming and gelling abilities. *European Food Research and Technology, 244*(1), 97–104.

Sun, H., Ju, Q., Ma, J., Chen, J., Li, Y., Yuan, Y., Hu, Y., Fujita, K., & Luan, G. (2019). The effects of extruded corn flour on rheological properties of wheat-based composite dough and the bread quality. *Food Science & Nutrition, 00*, 1–9.

Uysal, R. S., Mentes Yilmaz, O., & Boyaci, I. H. (2019). Determination of liquid egg composition using attenuated total reflectance Fourier transform infrared spectroscopy and chemometrics. *Journal of the Science of Food and Agriculture, 99*(7), 3572–3577.

Vegan Society of Aotearoa New Zealand. (2019). URL: http://www.vegansociety.org.nz/tryvegan/eggalternatives. Accessed on 28 Sept 2019.

Withana-Gamage, T. S., Wanasundara, J. P., Pietrasik, Z., & Shand, P. J. (2011). Physicochemical, thermal and functional characterisation of protein isolates from Kabuli and Desi chickpea (Cicer arietinum L.): A comparative study with soy (*Glycine max*) and pea (*Pisum sativum L.*). *Journal of the Science of Food and Agriculture, 91*(6), 1022–1031.

Zhang, B., Peng, H., Deng, Z., & Tsao, R. (2018). Phytochemicals of lentil (Lens culinaris) and their antioxidant and anti-inflammatory effects. *Journal of Food Bioactives, 1*, 93–103.

# Chapter 9
# Sprouting Water Composition

**Dan Xiong, Congyi Gao, Luca Serventi, Yuxin Cai, and Yaqi Bian**

## 9.1 Introduction

Legumes are plants in the family *Fabaceae* or Leguminosae. Common legumes include cultivated grains and beans such as soybeans, chickpeas, and lentils, etc. Legumes also have the ability to fix nitrogen in the atmosphere which is vital to environmental sustainability. The composition of legumes shows that they are the rich source of protein, vitamins, dietary fiber and minerals that can be beneficial to human health (Conde 2014). As for minerals, the contents of calcium and magnesium are generally high in legumes. A study found that soybeans deliver 201 and 220 mg/100 g of calcium and magnesium, respectively, while lentils have 7.5 mg of iron per 100 g of dry seed (Lin and Lai 2006). Apart from nutritive value, legumes contain antioxidant compounds including phenolic substances such as phenolic acids and lignin which potentially benefit human health (Lin and Lai 2006). Phenolic compounds are resistant to oxidation and prevent cells in tissue from degeneration and damage. It was also found that the phenolic content is heavily increased after sprouting in legumes (Khang et al. 2016). However, some antinutrients also exist in legumes such as lectin, phytic acid, polyphenols and saponins. These compounds limit the nutritional quality and affect the absorption of nutrients in human bodies. On this point, germination of legumes has been identified as an effective method to reduce the antinutrient factors and improve the legume quality.

Legume sprouts are also consumed as daily vegetables in many countries. Some studies have proved that germination of grains can significantly change the beneficial compounds by biochemical metabolisms. For instance, the activation of endoenzymes hydrolyses macronutrients including protein, starch and lipid. Sprouting of legumes also increase the nutrients level and have a positive effect on antioxidant

D. Xiong · C. Gao · L. Serventi (✉) · Y. Cai · Y. Bian
Department of Wine, Food and Molecular Biosciences, Faculty of Agriculture and Life Sciences, Lincoln University, Lincoln, Christchurch, New Zealand
e-mail: Luca.Serventi@lincoln.ac.nz



L. Serventi, *Upcycling Legume Water: from wastewater to food ingredients*,
https://doi.org/10.1007/978-3-030-42468-8_9

activity (Fouad and Rehab 2015). During sprouting, the vitamin content of vitamins of legumes increased significantly. For example, after 5 days of sprouting, vitamin C content of soybean sprouts increased from 2 to 11 mg/100 g (Ghani et al. 2016). Furthermore, a 4-day germination reduced by 71% the Kunitz Trypsin Inhibitor (KTI) (Kumar et al. 2019).

The objective of this book chapter is to assess the effect of sprouting on nutrients and antinutrients found in legumes, specifically minerals, vitamins, phytic acid and trypsin inhibitor. In addition, an experimental section analysed the wastewater from chickpea sprouting for mineral content and enzyme activity.

## 9.2 The Mechanism of Legume Sprouting

### *9.2.1 Physiological Factors*

Legume seeds start growing in the ovary and at maturity are composed of embryo surrounded by the endosperm. The major parts of legume seed are cotyledons, hypocotyl and seed coat. Cotyledons are the prominent parts, constituting around 90% of the whole seed. They are specialized seed leaves and develop from plumule, serving as storing-food tissue. The second biggest fraction is seed coat which consists of 9.9% of the whole dry mass (Moraghan et al. 2006). It is responsible for mechanical protection and transporting nutrients to the embryo and cotyledons. Seed coat also plays an important role in determining the development of seeds germination and dormancy.

Seeds are difficult to sprout without specific conditions. The main resistance factor is seeds dormancy. It is a kind of protection mechanism developed through genetic evolution that inhibits seeds sprouting in an unsuitable environment which may result in low incidence to survive, such as lack of moisture, oxygen and light, low or high temperature. Once the conditions are allowed, the specific genes will express in the seed coat tissues which trigger its action of germination immediately (Moïse et al. 2005). Regarding legume seeds, physical dormancy is normally founded, and it is also known as hardness. The seed coat becomes impermeable for water transition because of the palisade cell layers and the phenolics. In most cases, temperature is a key trigger for seeds becoming water-permeable, leading to the dormancy recovery and release of germination mechanisms (Smýkal et al. 2014).

Seeds sprouting start with water absorption, followed by the embryo expansion and a series of metabolic activations (Fig. 9.1). The third step involves in further uptake of water to help elongate the cells. This followed by the breaking through seed coat structure and endosperm rupture. The whole phase begins with genetic material transcription and translation. Once the radicle emerges, the cells start to divide, and the germination process is achieved (Koornneef et al. 2002).

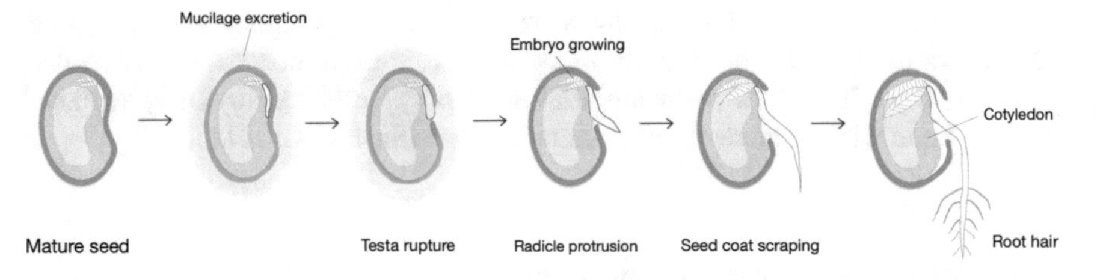


**Fig. 9.1** Representation of the germination progress of beans

**Table 9.1** Optimal temperatures for legume sprouting

| Legumes | Optimal temperature | References |
|---|---|---|
| Common beans | 22–31 °C | Neto et al. (2006) |
| Chickpeas | 20–29 °C | Soltani et al. (2006) |
| | 31.8–33 °C | Covell et al. (1986) |
| | 25–30 °C | Ellis et al. (1986) |
| Lentils | 24–24.4 °C | Covell et al. (1986) |
| | 20 °C | Rahimi et al. (2009) |
| Mung beans | 35 °C | Fyfield and Gregory (1989) |
| Peas | 20 °C | Raveneau et al. (2011) |
| Soybeans | 34–34.5 °C | Covell et al. (1986) |
| Faba bean | 25.5 °C | Dumur et al. (1990) |

### *9.2.2 Environmental Factors*

Under different conditions of free water availability and soil temperature, the rate of germination and seeds size may differ. The rate of sprouting is generally proportional to soil temperature in a certain range. Table 9.1 below shows the optimal temperature range for germination of different legumes. While the integral optimal temperature ranges from 20 to 34 °C, studies have shown that chickpea has the optimum germination temperature at 31–33 °C. Soybean germinates better at 34 °C (Covell et al. 1986) and mung bean has a higher optimal temperature at 40 °C (Kigel et al. 2015). As for water utilization, the sprouting rate and the volume of seed increase as the water availability increases within the thresholds. However, light is of less importance during the initial stage of sprouting.

The structure and hardness of seeds can influence sprouting. Therefore, in some food industrial processing, the raw seeds are normally softened before germination which can be achieved by soaking in the water or via thermal processing. Due to the thickness of legume seeds coat, may impact the permeability for exchanging nutrients from the external environment. On the other hand, antinutrients and toxic components can also be removed through thermal processing which would be more digestible for human consumption.

Many researches have proved that legume seeds components including water, macronutrients and minerals level change differently during germination. The moisture content and nutrients qualities are related to the germination time as well. One previous study of mung bean indicated that moisture content increased dramatically from 7% to 58% in the initial 24 hours of sprouting, followed by a gradual upward trend with an increase in germination time (Shah et al. 2011).

## 9.3 Macronutrients in Legumes

Energy is stored in most legume seeds in the form of some macronutrients, such as carbohydrates, lipids and proteins. These nutrients are mainly present in cotyledons and endosperms. The function is to provide food and release energy to embryo for the growth of the sprouts. Starch is the primary carbohydrate source which accounts for 45% in most legumes except soybeans (Lin and Lai 2006). Ghavidel and Prakash (2007) studied the effect of germination and dehulling on different legume seeds (mung beans, lentils, cowpea and chickpea) implicated that germination increased starch digestibility from 36% to 39%. This is partially because of the reduction of antinutrients level and amylase activity, the starch were hydrolysed into glucose and short chain polymers, leading to the decrease of total starch (Blessing and Gregory 2010). They also found the drop of starch content was followed by a steep increase of sugar. Martin-Cabrejas and collaborators (2008) suggested that the improvement of α-galactosidase activity caused break of α-1,6-galatosidic linkages that increase the sugar content. Whereas the sugar level appeared to decrease continuously till the end of germination. After these transformations, the germinated seeds were soft, easier to digest and even more palatable.

In terms of protein, they found its level increased in the range of 2–5% and the digestibility also improved remarkably by 13–16% (Ghavidel and Prakash 2007). Endosperm contain a big portion of protein of the whole seed, therefore dehulling can cause the increase of protein percentage. It is also related to the higher metabolic activity following soaking and germination that various carbohydrates are broken down by enzymes, as well as complex protein breakdown into digestible forms (Blessing and Gregory 2010). Another reason is that germination results in the reduction of phytic acid and tannin, which can bind with protein (Chaudhary et al. 2015).

Studies have shown that legumes are rich in dietary fibre content, mainly insoluble. The germination of legumes seed also have an impact on the dietary fibre fractions, mainly an increase. Though the its level would decrease remarkably after removing the hull, an increase level can be noted during later germination, due to the synthesis of structural carbohydrates contents including cellulose and hemicellulose, which are the main elements of cell wall. However, the extent of the change should be based on the sprouting conditions and legume variety. According to the findings of Martín-Cabrejas and collaborators (2008), the level of dietary fibre generally increased for the most of legumes except soybeans, which presented 20% of decrease. Whereas, in Lee's study of small black soybean, the germination process significantly increase the dietary fibre content (Lee et al. 2006).

Many research studies found a loss of lipid content after germination whereas the extent of the decrease depends on the type of legumes. Blessing and Gregory (2010) observed a slightly increase of lipid in mung bean after 24 hours of sprouting but following 59% loss in 72 hours. Maneemegalai and Nandakumar (2011) reported that after 72 hours germination, mung bean, black bean lost 17% and 11% of lipid, separately. According to Mubarak (2005) and Wilson's and Kwanyuen (1986) conclusions, the loss of lipid content of soybean is due to the use of lipid as carbon source for sprouting process and the hydrolysis of triacylglycerol degraded into fatty acid and glycerol.

## 9.4 Micronutrients in Legumes

### *9.4.1 Minerals*

Micronutrients are chemical substances required in trace amounts for the development of living systems. Legumes are a great natural source of micronutrients including minerals, vitamins and antioxidants. The main minerals in legume seeds are calcium, copper, iron, phosphorus and zinc. Minerals help the human body growth in various ways. For example, potassium and sodium can regulate our fluid balance, while calcium assists in building strong bone and teeth. Iron is the vital element for impairing the cognitive development in children and infant. Zinc has been reported as important to prevent diarrheal and respiratory diseases (Hotz and Brown 2004). Legumes are also a good source of folate which is a B-Vitamin and helps for megaloblastic anaemia treatment (Ministry of Health 2019). The ash contents account for small amount of proportion in legume seeds ranging from 2.4% to 4.6% in average (El Tinay et al. 1989). Most legumes contain the same variety of minerals while the proportion of different minerals may differ from legume species. For instance, calcium content is higher in soybean which is 287 mg/100 g. Lentil and soybean shows higher level of iron that is 10.2 mg and 11.7 mg per 100 g respectively. Common beans contain more sodium (36 mg/100 g) while others contain less than 20 mg (El Tinay et al. 1989). Some article indicated that mineral contents are also correlated with the weight of seeds. Meanwhile, the phosphorus level is positively correlated with the seed mass while the calcium level is negatively correlated with seed mass. As for potassium, zinc and iron, there are no significant correlation with the weight of seed (Moraghan and Grafton 2001). The distribution and accumulation of minerals in legume seeds is controlled by genetic factors. Study has found that most of the calcium is in the seed coat and more potassium, iron and zinc is located in embryo. Ribeiro and collaborators (2012) found that in common beans, over 95% of calcium is located on the seed coat, while 76–90% of potassium is found in the embryo. Therefore, it is recommended that beans are better consumed with seed coat otherwise the calcium content would remarkably decrease in meals. In addition, calcium cannot move freely between seed coat and embryo in bean seeds, probably because of the insoluble calcium oxalate crystals in mature seed coat (Barnabas and Arnott 1990). The concentrations of zinc, copper and iron vary heavily among embryo and seed coat fractions in different legume varieties (Ribeiro et al. 2012). Moraghan and Grafton observed black beans and found that the seed coat have 84% calcium, 24% iron, 8% potassium and 7% zinc of the entire seed (Moraghan and Grafton 2002). The mean zinc content in embryos ranged from 2.1 to 2.9 mg per 100 g dry beans and it is the highest proportion in embryos, ranging from 55% to 76%. The copper elements that distributed in the embryos range from 40% to 72% and in the seed coat from 28% to 60% (Ribeiro et al. 2012).

Studies have approved that sprouting of legumes improves the overall nutrients contents since the reduction of antinutritional factors. Minerals such as iron, calcium and zinc are released in soaking water or the formation of bound compounds. Figure 9.2 simulates this movement of calcium during the germination of chickpea. The availability of minerals is enhanced since the phytic acid decrease after germination (El-Adawy 2002). Based on different findings, the variation of mineral contents is different after germination. Bains and collaborators (2014) found no significant variation in iron and zinc content of cowpea and mung bean while chickpea had a significant reduction in iron level when soaked. However, El-Adawy (2002) studied chickpea seeds and found an increased trend of iron, phosphorus and zinc contents by 2%, 6% and 14%, respectively. In addition, the germinated chickpea showed considerable decreases in calcium, potassium, magnesium and copper contents. Same results observed in both Bains et al. (2014) and Barakoti and Bains' (2007) researches. They noted that calcium was low in germinated legumes when compared to raw seeds. In particular, the significant variation in calcium was found during soaking while germination had little effect on its level. Therefore, it is hypothesised that a fraction of minerals support the sprout growth, while another fraction could be leached in the sprouting water (Fig. 9.2). Section 9.6 addresses this hypothesis.

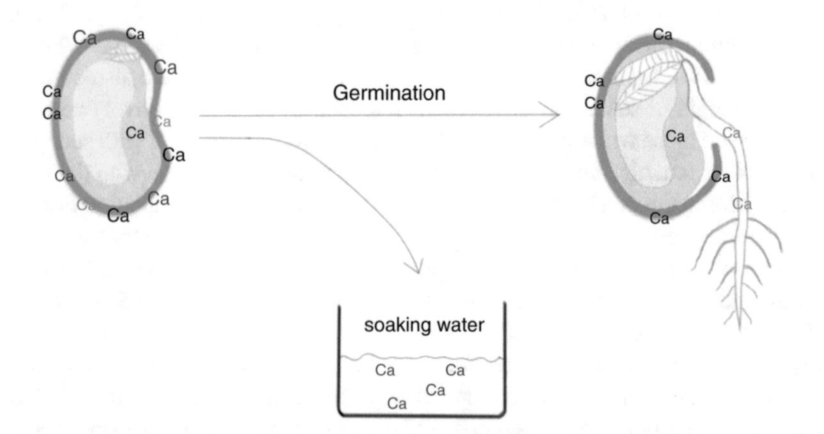


**Fig. 9.2** Simulated diagram of the mobilisation of calcium in chickpea seeds during germination

### 9.4.2 Phenolic Compounds

Sinapic acid and cinnamic acid are dominant phenolic components discovered in most legumes (Sebei et al. 2013). For common bean, some of the main phenolic components are flavonoids and tannins (black beans seed coat), polyphenols (extracted from pinto bean), anthocyanins (black beans, pink and pinto beans) (Beninger and Hosfield 2003). In addition, hydrolysed peptides also function as antioxidants since their amino acids contain phenolic, imidazole and indole groups which can react with free radicals (Oomah et al. 2010). Soluble phenolic compounds are abundant in soybeans. For instance, isoflavone and anthocyanin. Some methods showed that 90% of antioxidant properties come from the seed coat of soybeans. Particularly, black soybeans have been approved to have stronger radical scavenging activity than yellow and green soybeans (Kumar et al. 2010). Lentils and peas have different concentration of hydroxybenzoic phenolics including protocatechuic, vanillic acid and p-hydroxybenzoic (López-Amorós et al. 2006).

The sprouting process can alter the total phenolics level in legumes seeds. These antioxidants are naturally generated during the growth of sprouts to help them grow better and protect them from diseases and other environmental stress. The metabolism of germination starts in the presence of moisture, the seeds particles change and produce a large number of new components and energy, which include phenolics. Also, the various change in phenolics level not only depends on the type of legumes, but also the germination conditions such as light existence and duration of sprouting (López-Amorós et al. 2006). Khang and collaborators (2016) found that the phenolic contents increased twofold after germination for 5 days. They quantified thirteen phenolic compounds in different legumes including p-hydroxybenzoic acid, protocatechuic acid, caffeic acid, vanillin, and more, then detected nine in black beans, white beans and mung beans, ten in soybeans and seven kinds in peanuts. The legume sprouts samples all obtain sinapic acid and cinnamic acid, and the results of mean phenolic content show in Table 9.2 below. They concluded that germinated soybeans and peanuts contained the highest level of antioxidant capacity and phenolics. For common beans, they generated benzoic acid the most after germination (Khang et al. 2016).

Similar changes were observed by López-Amorós and co-authors (2006). They found that during soaking, the phenolic contents all decreased in lentils, beans and

**Table 9.2** Total mean phenolic content of raw and germinated legumes (mg GAE /g dry weight) (Khang et al. 2016)

| Time | Soybeans | Cowpeas | Black beans | Mung beans | Peanuts |
|---|---|---|---|---|---|
| 0 hours | 12.1 | 7.8 | 11.7 | 5.8 | 18.2 |
| 24 hours | 21.5 | 11.3 | 13.2 | 11.5 | 26.9 |
| 48 hours | 21.6 | 14.1 | 13.6 | 12.3 | 28.7 |
| 72 hours | 23.9 | 14.7 | 14.2 | 12.3 | 29.4 |
| 96 hours | 24.8 | 16.3 | 13.5 | 12.5 | 30.3 |
| 120 hours | 28.3 | 19.5 | 16.5 | 15.0 | 37.6 |

peas, along with a later increasing trend in germination period. The total antioxidant activity is correlated to time and light. In terms of beans, the highest activity was observed after 6 days of sprouting without the light. For peas, it reached peak value after 4 days of sprouting in the absence of light. Particularly, the p-hydroxybenzoic aldehyde and vanillic aldehyde were found in germinated beans and lentils, with a higher concentration in the absence of sunlight. The differences may be explained by the synergetic effects by different bioactive substances except polyphenols such as vitamins and carotenoids. However, the germinated lentils were detected to have lower antioxidant capacity than raw ones which showed a opposite tendency compared to beans and peas.

### *9.4.3 Vitamins*

Vitamins B are a group water-soluble vitamins including B1 (thiamine), B2 (riboflavin), B3 (niacin), and B6 (pyridoxine) and others (Mozafar 2018; Winkler et al. 2006). Plaza and collaborators (2003) compared the vitamin content of soybeans and alfalfa between sprouts and dry seeds after 4 days sprouting in the darkness at 28 °C. In this experiment, all of the vitamin B complexes of soybean and alfalfa increased. For soybean, after 4 days of sprouting, the vitamin B2 (riboflavin) content increased more than tenfolds. Soybean sprouts contained 2 times more riboflavin than alfalfa sprouts. The mechanisms for this increase are currently unknown. The increase in thiamine was quantified from 8.8 to 19.6 μg/100 g, whereas, the increase in niacin and riboflavin were from 21.4 to 64.4 μg/100 g and from two to tenfolds, respectively (Table 9.3).

Vitamin C is another water-soluble vitamin. It is also called ascorbic acid and is essential organic compounds on human health because humans cannot synthesize vitamins by themselves (Ghani et al. 2016). Several experiments have proved that sprouting can increase vitamin C content. Plaza and collaborators (2003) tested the differences of vitamin C content between dry seed with sprouts of alfalfa and soybean at 28 °C 4 days, showing ascorbic acid increase over time (Table 9.3). The ascorbic acid content increased by about tenfolds, from 81 to 827 μg/g, in alfalfa sprouts, whereas by about threefolds, from 100 to 316 μg/g, in soybean sprouts (Plaza et al. 2003).

Devi and coauthors (2015) sprouted cowpea seeds at 25 °C for 24 hours. After sprouting, the vitamin C content in cowpea seed increased from 0.33 to 2.7 mg/100 g. Islam and co-authors (2017) tested the vitamin C content in mung bean and Lignosus bean. They did not detect vitamin C in dry mung bean and Lignosus bean. However, during 3 and 5 days sprouting at 25 °C and 30 °C of mung bean sprouts and Lignosus bean sprouts, vitamin C was detected in the amounts of 9.5 and 21 mg/100 g. Lien and others (2016) also tested the vitamin C content change during germination. In their experiments, soybeans germinated at 25 °C for 72 hours had the highest vitamin C content, having increased from 4.7 to 13 mg/g (Table 9.3).

**Table 9.3** Changes in vitamin content of representative legumes after sprouting

| Vitamin | Legume | Change (μg/100 g) | References |
|---|---|---|---|
| A | Alfalfa | From 0.42 to 532 | Plaza et al. (2003) |
| | Soybean | From 19 to 34 | Plaza et al. (2003) |
| | | From 6.6 to 33 | Kang et al. (2012) |
| B1 | Mung bean | From 8.8 to 8.2 | Burkholder and McVeigh (1945) |
| | Soybean | From 16 to 20 | Burkholder and McVeigh (1945) |
| B2 | Alfalfa | From 169 to 698 | Plaza et al. (2003) |
| | Soybean | From 129 to 1480 | Plaza et al. (2003) |
| B3 | Mung bean | From 23 to 64 | Burkholder and McVeigh (1945) |
| | Soybean | From 21 to 30 | Burkholder and McVeigh (1945) |
| C | Alfalfa | From 81 to 827 | Plaza et al. (2003) |
| | Cowpea | From 0.33 to 2.7 | Devi et al. (2015) |
| | Lignosus bean | From 0 to 20,940 | Islam et al. (2017) |
| | Mung bean | From 0 to 9.5 | Islam et al. (2017) |
| | Soybean | From 100 to 316 | Plaza et al. (2003) |
| E | Alfalfa | From 10 to 38 | Plaza et al. (2003) |
| | Soybean | From 80 to 250 | Lien et al. (2016) |

The reasons of vitamin C increase could be the glucuronate pathway. In plants, l-GalL dehydrogenase (GalLDH, EC 1.3.2.3) may catalyze the oxidation of l-GalL dehydrogenase to AsA (L-ascorbic acid or vitamin C) in the final step of L-ascorbic acid biosynthesis (Tabata et al. 2001). The experiment of Isherwood and co-authors (1954) showed that L-gulOnO-, L-galactono- and D-glucurono-y-lactones and D-galacturonic acid methyl ester could be transformed into L-ascorbic acid. These four compounds are intermediates of reactions between D-glucose and L-ascorbic acid. Vitamin C is produced with the cells, across all seed. Above results show that sprouting can increase the content of vitamin C, quantified in values from not detectable to 20.9 mg/100 g in some legume sprouts, such as cowpea sprouts, mung bean sprouts and lignosus bean sprouts. However, in soybean sprouts and alfalfa sprouts, the increases of content of ascorbic acid was from 81 to 827 μg/g.

Apart from water-soluble vitamins such as vitamin B and vitamin C, legume sprouts can also contain other fat-soluble vitamins. The main fat-soluble vitamins are vitamin A (retinol) and vitamin E (α-tocopherol). Plaza and collaborators (2003) tested soybean and alfalfa sprouts and seed vitamins content. The results of this experiment show the vitamin A content in alfalfa from dry seed to sprouts increased from 0.42 to 53 RE/100 g during 4 days sprouting. At the same time, the vitamin A content in soybean increased twofolds from 19 to 34 RE/100 g. The vitamin E content in alfalfa increased from 10 to 38 μg/g and the vitamin E content in soybean increased from 0.89 to 34 μg/g. Compared with water-soluble vitamin content, alfalfa had higher content in fat-soluble vitamins. In general, the fat-soluble vitamins are affected by sprouting to lower extent than water-soluble vitamins. Lien and co-authors (2016) analysed legume seeds germinated for 72 hours across a range of temperatures: 22–28 °C. The highest vitamin E content in soybean was the sample germinated at 25 °C which increased from 0.08 to 0.25 mg/g. The β-carotene content in soybean had similarly increased. For instance, during 5 days sprouting, β-carotene content in soybean sprouts are increased by 33.3 μg/g between soybean seed was 6.6 μg/g (Kang et al. 2012). The increases of content of vitamin A was from 0.42 to 532 μg/100 g, while changes from 0.08 to 38 μg/100 g were observed for vitamin E. The results described above show that sprouting can increase the content of varieties of vitamin A and vitamin E.

## 9.5 Antinutrients in Legumes

### *9.5.1 Phytic Acid*

Although legumes seeds are regarded as good source of nutritive values, there are a range of bioactive compounds that cannot be defined as nutrients, including lectins, phytic acid, tannins, proteolytic inhibitors, alkaloids, saponins, isoflavones, oligosaccharides etc. These compounds can be either toxic or inhibiting of metabolism in the human body (Lal et al. 2017). Abundant studied have approved that germination have an effect on antinutrients profile. Phytic acid is the primary phosphorus contents in legumes and one of the issues is that it have the great potential to combine with minerals, resulting in the formation of insoluble mineral-phytic acid molecules which are unavailable for human beings and thus, cause mineral ions deficiency (Bau et al. 1997).

A number of researchers have observed a decrease in phytic acid after sprouting due to a large increase in phytase activity. Sinha and Kawatra (2003) reported that phytic acid of cowpea decreased by 16% and 30% during soaking and dehulling, respectively. Ghavidel and Prakash (2007) reported that germination caused a drastic decrease of tannin and phytic acid levels by 43–52% and 47–52%, respectively. They also found a significant negative correlations between nutrients bioavailability and anti-nutrients factors. In other words, dehulling and germination process could reduce the anti-nutrients level and enhance the digestibility of nutrients in legume seeds. The extent of decrease can be determined by germination time and legume varieties. Bains and collaborator (2014) found that the reduction in phytic acid increased with the sprouting duration. Moreover, Egli and co-authors (2002) noted that phytase activity in chickpea and bean remained steady within the first day, but increased thereafter. In Table 9.4, Mittal and co-authors (2012) reported 93% and

**Table 9.4** Reduction (percentage) of antinutrients in chickpeas after germination (Mittal et al. 2012; Singh et al. 2015)

| Antinutrient | Percent reduction |
|---|---|
| Phytic acid | 3% |
| Trypsin inhibitor | 25–85% |
| Tannins | 93% |
| Polyphenols | 82% |

82% decrease in tannin and polyphenols, separately after 48-hour germination. Legumes seeds generally have more than 20% of oligosaccharides contents such as raffinose and stachyose, which can cause flatulence in animals. Since these components provide as carbon sources during sprouting, their levels would reduce after germination.

### 9.5.2 *Trypsin Inhibitor*

Table 9.4 shows the change in trypsin inhibitor activity (TIA) from chickpeas after the germination for 72 hours. Some of the cultivars of do not a significant reduction during 3 days sprouting. For example, the trypsin inhibitor activity of the BG 391 cultivar decreased from 0.18 to 0.13 mg/g (Singh et al. 2015). In addition, trypsin inhibitor activity of the JG63 reduced from 0.12 to 0.09 mg/g. However, the trypsin inhibitor activity of some of them has a huge reduction. TIA of JG130 decreased to 43%, from 0.28 to 0.12 mg/g. At the same time, JG74 chickpea's TIA reduced from 0.14 to 0.02 mg/g (Singh et al. 2015). The above results show sprouting can decrease the content of varieties of TIA.

## 9.6 Minerals and Enzymes in Sprouting Water

### 9.6.1 *Introduction*

Significant changes in minerals and enzymes were observed upon germination of legume seeds (Bains et al. 2014; Barakoti and Bains 2007; Blessing and Gregory 2010). Therefore, an experiment was performed to determine whether the wastewater produced by legume sprouting contains any nutrient.

### 9.6.2 *Materials and Methods*

A sprouter (Mr. Fothergill's Kitchen Seed Sprouter, New Zealand) was used to germinate 50 g of soaked chickpeas, spread evenly over four plastic trays. The chickpeas have been previously soaked overnight with a 3.3 to 1 ratio of water to legume. The trays were stacked on top of each other, above a collecting bottom. Each tray was equipped with an overflow valve (water outlet) that allowed draining of water to the lower tray. Water was added to cover the water outlet, in the amount of 200 ml, two times per day, over a 2-day period, for a total of 400 ml per day. Aliquots of sprouting wastewater were collected from the bottom tray on both days, keeping them separate. Mineral content of the freeze-dried samples was determined with an official method (AOAC, 930.05, 1995). A kit for colorimetric assay of enzymes cellulase and protease was supplied by GlycoSpot (Søborg, Denmark). Two substrate plates were used: cellulose-based for cellulase and casein-based for protease. Activation of the substrate plate was achieved with 200 μl of solution added to each well, followed by incubation for 10 minutes at room temperature without agitation. Samples were centrifuged (Heraeus Multifuge XIR, Thermo Scientific, MA, USA) for 10 minutes at 2700 g. The activation solution was removed by washing twice with 100 μl of distilled water, then a buffer was added to each well, followed by samples and control (water). A product plate was located underneath the substrate plate. After incubating and shaking, the product plate was centrifuged (Heraeus Multifuge XIR, Thermo Scientific, MA, USA) for 10 minutes at 2700 g. Finally, absorbance was measured at 517 nm (protease) and 595 nm (cellulase) with a FLUOstar Omega microplate reader (BMG Labtech, New Zealand) and results expresses as optical density (OD).

### *9.6.3 Results and Discussion*

Both samples collected (day 1 and day 2) contained about 1 g/100 g of dry matter. The freeze-dried powders of chickpea sprouting water were found to contain minerals. The samples collected after 1 day of germination contained 30.2 g/100 g of minerals, while sample from day 2 averaged 34.9 g/100 g of minerals (Table 9.5). A previous study (Erba et al. 2019) has shown that sprouted chickpeas contain significantly less minerals than the raw ones: 26 vs. 28 g/kg dry matter. The main loss was represented by calcium: 545 vs. 791 mg/kg dry matter (Erba et al. 2019). The same study also indicated higher protein content (202 vs. 186 g/kg dry matter) (Erba et al. 2019). This means that about 0.2–0.3 g of minerals are lost for every 100 g of chickpeas germinated. In this study, 50 g of chickpeas released 0.30–0.35 g of minerals each day, in agreement with literature. Therefore, these preliminary findings may indicate that some of the mineral loss is due to leaching in the processing water.

The qualitative analysis of enzymes revealed interesting results. No cellulase was found, while a significant protease activity was detected, specifically on day 2 (Table 9.5). A study determined that two-third of seed germination was stimulated by a chemical signal, which may derive from hemicellulose degradation (Smith and Van Staden 1995). It is therefore possible that the 2-day germination was not sufficient to activate cellulase enzymes or, in case it was, no leaching occurred. On the

**Table 9.5** Mineral and enzyme content of chickpea sprouting water collected at Day 1 and Day 2 of germination

| Nutrient | Day 1 | Day 2 |
|---|---|---|
| Minerals (g/100 g dry matter) | 30.2 | 34.9 |
| Enzyme cellulase (OD) | <dl | <dl |
| Enzyme protease (OD) | 0.14 | 0.42 |

The term < means below detection limit

opposite, for the protease, a large increase from day 1 to day 2 was observed. The protease content in day 2 sprouting water was 0.42 OD, increasing drastically from 0.14 OD of day 1. That means proteolytic enzymes have been released during germination and subsequently leached in the wastewater. Previous studies have shown increased amino acid content in legume sprouts (Erba et al. 2019; Masood et al. 2014). Specifically, protein degradation was observed, as resulted of protease activity during germination, which depolymerised storage proteins to release amino acids, supporting structural growth of the sprouts (Masood et al. 2014). Consequently, this experiment supports the hypothesis that at least some of the proteolytic enzymes are leached in the wastewater during sprouting of legumes.

## 9.7 Conclusions

Legumes are rich in protein, dietary fibre, minerals and phytochemicals. It is known that germination is a good way to enhance nutritional quality and reduce antinutritional contents, because it can result in alteration of the composition of these materials and improvement in digestibility. Therefore, the understanding of sprouting mechanism of legumes and how exactly nutrients quality change after sprouting is significantly necessary. The two major factors that induce germination are water and appropriate temperature for certain legume. Every legume has its optimal temperature for sprouting. However, other agreed conclusions are germination can increase the total phenolics level and some of the macronutrients in legume seeds. Most researches recorded that germination improves the starch digestibility, protein and fibre level, whereas lipids and sugar content appeared to drop. In addition, the sprouting process increased the content of a variety of vitamins. The content of the thiamine of soybean sprouts increased from 16 to 20 μg/g. At the same time, the content of the niacin of soybean sprouts also increased from 21 to 29.9 μg. Compared with dry soybean seeds, the content of riboflavin of soybean sprouts increased two-folds. In addition to various vitamin B, after germination, the content of vitamin C, A and E are also increased in legume sprouts. For example, in soybean sprouts, the content of vitamin C increased from 81 to 827 μg/g. Meanwhile, the content of vitamin A also increased from 19 to 34 mg/100 g. Meanwhile, the content of vitamin E increased from 0.8 to 34 μg/g. Sprouting can decrease TIA of legumes even though to a lower extent than for boiling.

As for antinutrients such as phytic acid and trypsin inhibitor, most studies agreed that germination could be the useful and effective way to improve the nutritional quality of legumes. Moreover, some of the nutrients are lost in the sprouting process such as minerals, phenolics and oligosaccharide. A recent experiment has proven that 30–35% of the solid loss in sprouting water was minerals. Further studies are warranted to identify the mineral profile as well as the remaining solid fraction. Furthermore, high levels of protease activity were detected in the sprouting water, from 0.14 to 0.42 OD in day 1 and 2.

These findings suggest potential for legume sprouting water as source of protease. In closing, sprouting water can be a useful by-product for food applications as texturizers and source of nutrients.

**Acknowledgments** The realisation of this book chapter (literature review and experimental data) was made possible by the teaching funding allotted by Lincoln University (New Zealand) to the bachelor courses named "FOOD 398 – Research Essay" and "FOOD 399 – Research Placement". Authors would like to thank Letitia Stipkovits for supporting the enzymatic analysis of sprouting water

## References

AOAC. (1995). *Official methods of analysis of AOAC international* (16th ed.). Aarlington: AOAC International Publication.

Bains, K., Uppal, V., & Kaur, H. (2014). Optimization of germination time and heat treatments for enhanced availability of minerals from leguminous sprouts. *Journal of Food Science and Technology, 51*(5), 1016–1020.

Barakoti, L., & Bains, K. (2007). Effect of household processing on the in vitro bioavailability of iron in mungbean (Vigna radiata). *Food and nutrition bulletin, 28*(1), 18–22.

Barnabas, A. D., & Arnott, H. J. (1990). Calcium oxalate crystal formation in the bean (Phaseolus vulgaris L.) seed coat. *Botanical Gazette, 151*(3), 331–341.

Bau, H. M., Villaume, C., Nicolas, J. P., & Méjean, L. (1997). Effect of germination on chemical composition, biochemical constituents and antinutritional factors of soya bean (Glycine max) seeds. *Journal of the Science of Food and Agriculture, 73*(1), 1–9.

Beninger, C. W., & Hosfield, G. L. (2003). Antioxidant activity of extracts, condensed tannin fractions, and pure flavonoids from Phaseolus vulgaris L. seed coat color genotypes. *Journal of Agricultural and Food Chemistry, 51*(27), 7879–7883.

Blessing, I. A., & Gregory, I. O. (2010). Effect of processing on the proximate composition of the dehulled and undehulled mungbean [*Vigna radiata* (L.) Wilczek] flours. *Pakistan Journal of Nutrition, 9*(10), 1006–1016.

Burkholder, P. R., & McVeigh, I. (1945). Vitamin content of some mature and germinated legume seeds. *Plant Physiology, 20*(2), 301.

Chaudhary, R., Oluyemisi, A. E., Shrestha, A. K., & Adhikari, B. M. (2015). Effect of germination on biochemical and nutritional quality of K wati. *Journal of Food Processing and Preservation, 39*(6), 1509–1517.

Conde, N. (2014). Nutrition facts for Chickpeas (garbanzo beans, bengal gram), mature seeds, cooked, boiled, without salt. Retrieved from http://nutritiondata.self.com/facts/legumes-and-legume-products/4326/2.

Covell, S., Ellis, R. H., Roberts, E. H., & Summerfield, R. J. (1986). The influence of temperature on seed germination rate in grain legumes: I. A comparison of chickpea, lentil, soyabean and cowpea at constant temperatures. *Journal of Experimental Botany, 37*(5), 705–715.

Devi, C. B., Kushwaha, A., & Kumar, A. (2015). Sprouting characteristics and associated changes in nutritional composition of cowpea (Vigna unguiculata). *Journal of food science and technology, 52*(10), 6821–6827.

Dumur, D., Pilbeam, C. J., & Craigon, J. (1990). Use of the Weibull function to calculate cardinal temperatures in faba bean. *Journal of Experimental Botany, 41*(11), 1423–1430.

Egli, I., Davidsson, L., Juillerat, M. A., Barclay, D., & Hurrell, R. F. (2002). The influence of soaking and germination on the phytase activity and phytic acid content of grains and seeds potentially useful for complementary feedin. *Journal of Food Science, 67*(9), 3484–3488.

El Tinay, A. H., Mahgoub, S. O., Mohamed, B. E., & Hamad, M. A. (1989). Proximate composition and mineral and phytate contents of legumes grown in Sudan. *Journal of Food Composition and Analysis, 2*(1), 69–78.

El-Adawy, T. A. (2002). Nutritional composition and antinutritional factors of chickpeas (*Cicer arietinum L.*) undergoing different cooking methods and germination. *Plant Foods for Human Nutrition, 57*(1), 83–97.

Ellis, R. H., Covell, S., Roberts, E. H., & Summerfield, R. J. (1986). The influence of temperature on seed germination rate in grain legumes: II. Intraspecific variation in chickpea (*Cicer arietinum L.*) at constant temperatures. *Journal of Experimental Botany, 37*(10), 1503–1515.

Erba, D., Angelino, D., Marti, A., Manini, F., Faoro, F., Morreale, F., Pellegrini, N., & Casiraghi, M. C. (2019). Effect of sprouting on nutritional quality of pulses. *International Journal of Food Sciences and Nutrition, 70*(1), 30–40.

Fouad, A. A., & Rehab, F. M. (2015). Effect of germination time on proximate analysis, bioactive compounds and antioxidant activity of lentil (*Lens culinaris Medik.*) sprouts. *Acta Scientiarum Polonorum. Technologia Alimentaria, 14*(3), 233–246.

Fyfield, T. P., & Gregory, P. J. (1989). Effects of temperature and water potential on germination, radicle elongation and emergence of mungbean. *Journal of Experimental Botany, 40*(6), 667–674.

Ghani, M., Kulkarni, K. P., Song, J. T., Shannon, J. G., & Lee, J. D. (2016). Soybean sprouts: A review of nutrient composition, health benefits and genetic variation. *Plant Breeding and Biotechnology, 4*(4), 398–412.

Ghavidel, R. A., & Prakash, J. (2007). The impact of germination and dehulling on nutrients, antinutrients, in vitro iron and calcium bioavailability and in vitro starch and protein digestibility of some legume seeds. *LWT-Food Science and Technology, 40*(7), 1292–1299.

Hotz, C., & Brown, K. H. (2004). Assessment of the risk of zinc deficiency in populations and options for its control. International Nutrition Foundation for The United Nations University, Tokyo, Japan.

Kang, E. Y., Kim, E. H., Chung, I. M., & Ahn, J. K. (2012). Variation of β-carotene concentration in soybean seed and sprout. *Korean Journal of Crop Science, 57*(4), 324–330.

Khang, D., Dung, T., Elzaawely, A., & Xuan, T. (2016). Phenolic profiles and antioxidant activity of germinated legumes. *Food, 5*(2), 27.

Kigel, J., Rosental, L., & Fait, A. (2015). Seed physiology and germination of grain legumes. In *Grain legumes* (pp. 327–363). New York: Springer.

Koornneef, M., Bentsink, L., & Hilhorst, H. (2002). Seed dormancy and germination. *Current Opinion in Plant Biology, 5*(1), 33–36.

Kumar, V., Rani, A., Dixit, A. K., Pratap, D., & Bhatnagar, D. (2010). A comparative assessment of total phenolic content, ferric reducing-anti-oxidative power, free radical-scavenging activity, vitamin C and isoflavones content in soybean with varying seed coat colour. *Food Research International, 43*(1), 323–328.

Kumar, V., Rani, A., Mittal, P., & Shuaib, M. (2019). Kunitz trypsin inhibitor in soybean: Contribution to total trypsin inhibitor activity as a function of genotype and fate during processing. *Journal of Food Measurement and Characterization, 13*(2), 1583–1590.

Lal, N., Barcchiya, J., Raypuriya, N., & Shiurkar, G. (2017). Anti-nutrition in legumes: Effect in human health and its elimination. *Innovative Farming, 2*(1), 32–36.

Lee, C. H., Oh, S. H., Yang, E. J., & Kim, Y. S. (2006). Effects of raw, cooked, and germinated small black soybean powders on dietary fiber content and gastrointestinal functions. *Food Science and Biotechnology, 15*(4), 635–638.

Lien, D. T. P., Phuc, T. M., Tram, P. T. B., & Toan, H. T. (2016). Time and temperature dependence of antioxidant activity from soybean seeds (*Glycine max L. Merr.*) during germination. *Intern. J. Food Sci. Nutr, 1*, 22–27.

Lin, P. Y., & Lai, H. M. (2006). Bioactive compounds in legumes and their germinated products. *Journal of Agricultural and Food Chemistry, 54*(11), 3807–3814.

López-Amorós, M. L., Hernández, T., & Estrella, I. (2006). Effect of germination on legume phenolic compounds and their antioxidant activity. *Journal of Food Composition and Analysis, 19*(4), 277–283.

Maneemegalai, S., & Nandakumar, S. (2011). Biochemical studies on the germinated seeds of Vigna radiata (L.) R. Wilczek, Vigna mungo (L.) Hepper and Pennisetum typhoides (Burm f.) Stapf and CE Hubb. *International Journal of Agricultural Research, 6*(7), 601–606.

Martín-Cabrejas, M. A., Díaz, M. F., Aguilera, Y., Benítez, V., Mollá, E., & Esteban, R. M. (2008). Influence of germination on the soluble carbohydrates and dietary fibre fractions in non-conventional legumes. *Food Chemistry, 107*(3), 1045–1052.

Masood, T., Shah, H. U., & Zeb, A. (2014). Effect of sprouting time on proximate composition and ascorbic acid level of mung bean (*Vigna radiate L.*) and chickpea (*Cicer Arietinum L.*) seeds. *The Journal of Animal & Plant Sciences, 24*(3), 850–859.

Ministry of Health. (2019). Folate/folic acid. Retrieved from https://www.health.govt.nz/our-work/preventative-health-wellness/nutrition/folate-folic-acid.

Mittal, R., Nagi, H. P. S., Sharma, P., & Sharma, S. (2012). Effect of processing on chemical composition and antinutritional factors in chickpea flour. *Journal of Food Science and Engineering, 2*(3), 180.

Moïse, J. A., Han, S., Gudynaitę-Savitch, L., Johnson, D. A., & Miki, B. L. (2005). Seed coats: Structure, development, composition, and biotechnology. *Vitro Cellular & Developmental Biology-Plant, 41*(5), 620–644.

Moraghan, J. T., & Grafton, K. (2001). Genetic diversity and mineral composition of common bean seed. *Journal of the Science of Food and Agriculture, 81*(4), 404–408.

Moraghan, J. T., & Grafton, K. (2002). Distribution of selected elements between the seed coat and embryo of two black bean cultivars. *Journal of Plant Nutrition, 25*(1), 169–176.

Moraghan, J. T., Etchevers, J. D., & Padilla, J. (2006). Contrasting accumulations of calcium and magnesium in seed coats and embryos of common bean and soybean. *Food Chemistry, 95*(4), 554–561.

Mozafar, A. (2018). *Plant vitamins*. CRC Press, Boca Raton, FL, USA.

Mubarak, A. E. (2005). Nutritional composition and antinutritional factors of mung bean seeds (Phaseolus aureus) as affected by some home traditional processes. *Food Chemistry, 89*(4), 489–495.

Neto, N. B. M., Prioli, M. R., Gatti, A. B., & Cardoso, V. J. M. (2006). Temperature effects on seed germination in races of common beans (*Phaseolus vulgaris L.*). *Acta Scientiarum. Agronomy, 28*(2), 155–164.

Oomah, B. D., Corbé, A., & Balasubramanian, P. (2010). Antioxidant and anti-inflammatory activities of bean (*Phaseolus vulgaris L.*) hulls. *Journal of Agricultural and Food Chemistry, 58*(14), 8225–8230.

Plaza, L., de Ancos, B., & Cano, P. M. (2003). Nutritional and health-related compounds in sprouts and seeds of soybean (*Glycine max*), wheat (*Triticum aestivum. L*) and alfalfa (*Medicago sativa*) treated by a new drying method. *European Food Research and Technology, 216*(2), 138–144.

Rahimi, A., Norton, R., McNeil, D., & Hoseini, S. M. (2009). Effects of salinity and temperature on germination, seedling growth and ion relations of two lentil (*Lens culinaris*) cultivars. *Seed Technology, 31*(1), 76–86.

Raveneau, M. P., Coste, F., Moreau-Valancogne, P., Lejeune-Hénaut, I., & Durr, C. (2011). Pea and bean germination and seedling responses to temperature and water potential. *Seed Science Research, 21*(3), 205–213.

Ribeiro, N. D., Maziero, S. M., Prigol, M., Nogueira, C. W., Rosa, D. P., & Possobom, M. T. D. F. (2012). Mineral concentrations in the embryo and seed coat of common bean cultivars. *Journal of Food Composition and Analysis, 26*(1–2), 89–95.

Sebei, K., Gnouma, A., Herchi, W., Sakouhi, F., & Boukhchina, S. (2013). Lipids, proteins, phenolic composition, antioxidant and antibacterial activities of seeds of peanuts (*Arachis hypogaea l*) cultivated in Tunisia. *Biological Research, 46*(3), 257–263.

Shah, S. A., Zeb, A., Masood, T., Noreen, N., Abbas, S. J., Samiullah, M., Alim, M. A., & Muhammad, A. (2011). Effects of sprouting time on biochemical and nutritional qualities of Mungbean varieties. *African Journal of Agricultural Research, 6*(22), 5091–5098.

Sinha, R., & Kawatra, A. (2003). Effect of processing on phytic acid and polyphenol contents of cowpeas [Vigna unguiculata (L) Walp]. *Plant Foods for Human Nutrition, 58*(3), 1–8.

Singh, P. K., Shrivastava, N., Sharma, B., & Bhagyawant, S. S. (2015). Effect of domestic processes on chickpea seeds for antinutritional contents and their divergence. *American Journal of Food Science and Technology, 3*(4), 111–117.

Smith, M., & Van Staden, J. (1995). Infochemicals: The seed-fungus-root continuum. A review. *Environmental and Experimental Botany, 35*(2), 113–123.

Smýkal, P., Vernoud, V., Blair, M. W., Soukup, A., & Thompson, R. D. (2014). The role of the testa during development and in establishment of dormancy of the legume seed. *Frontiers in Plant Science, 5*, 351.

Soltani, A., Robertson, M. J., Torabi, B., Yousefi-Daz, M., & Sarparast, R. (2006). Modelling seedling emergence in chickpea as influenced by temperature and sowing depth. *Agricultural and Forest Meteorology, 138*(1–4), 156–167.

Tabata, K., Ôba, K., Suzuki, K., & Esaka, M. (2001). Generation and properties of ascorbic acid-deficient transgenic tobacco cells expressing antisense RNA for L-galactono-1, 4-lactone dehydrogenase. *The Plant Journal, 27*(2), 139–148.

Wilson, R. F., & Kwanyuen, P. (1986). Triacylglycerol synthesis and metabolism in germinating soybean cotyledons. *Biochimica et Biophysica Acta (BBA)-Lipids and Lipid Metabolism, 877*(2), 231–237.

Winkler, C., Wirleitner, B., Schroecksnadel, K., Schennach, H., & Fuchs, D. (2006). Beer down-regulates activated peripheral blood mononuclear cells in vitro. *International Immunopharmacology, 6*(3), 390–395.

# Chapter 10
# Bioactives in Legumes

**Luca Serventi and Lirisha Vinola Dsouza**

## 10.1 Bioactives from Legumes

Legumes are among the most widely cultivated crops in the world. They belong to the *Leguminosae* family and are important food sources, providing great beneficial health benefits due to their diverse nutritional composition. Legumes are consumed as basic staple food in many developing countries and were previously known as the "poor man's meat" in ancient Europe. Different types of legumes are consumed by humans: beans, chickpeas, cowpeas, green lentils, pigeon peas, soybeans and many more. In recent years, several studies showed the potential health benefits of legumes beyond satisfying the basic nutrient requirements for humans (Barman et al. 2018; de Cedrón et al. 2018; Ganesan and Xu 2017; Papandreou et al. 2019; Sathoff and Samac 2019; Torres-Fuentes et al. 2015; Zhang et al. 2018). The availability of numerous bioactive compounds in legumes is associated with many health promoting factors and prevention of various types of cancer (Roy et al. 2010) (Table 10.1).

Bioactive peptides are a group of proteins with short molecular chain (3–20 amino acids) (Moughan 2009) that occur naturally or may be derived from enzymatic hydrolysis of proteins. Numerous bioactives have been identified in legumes, expressing different functional significance (Malaguti et al. 2014). Depending upon the amino acid sequence present, bioactive peptides from legumes showed to have a major influence on the human body, performing different activities such as antimicrobial, blood-pressure-lowering, antioxidant and enhancement of mineral and vitamin absorption (Margier et al. 2018). The high percentage of bioactives in legumes results in increased the protein quality in food (Singh et al. 2014).

Notably, germination is an effective way to increase the amount of bioactives present in legumes. Bioactives compounds consist of phenolic acids, flavonoids,

L. Serventi (✉) · L. V. Dsouza
Department of Wine, Food and Molecular Biosciences, Faculty of Agriculture and Life Sciences, Lincoln University, Lincoln, Christchurch, New Zealand
e-mail: Luca.Serventi@lincoln.ac.nz



L. Serventi, *Upcycling Legume Water: from wastewater to food ingredients*,
https://doi.org/10.1007/978-3-030-42468-8_10

**Table 10.1** Bioactives from legumes and their main activities in humans (Kamran and Reddy 2018; Luna-Vital and de Mejía 2018; Malaguti et al. 2014; Reyes-Díaz et al. 2019; Sathoff and Samac 2019)

| Phenolics | Peptides | Saponins | Vitamins | Lipids |
|---|---|---|---|---|
| Anticancer | Anticancer | Anticancer | Anti-inflammatory | Anticancer |
| Anti-inflammatory | Anti-inflammatory | Anti-obesity | Antioxidant | Anti-inflammatory |
| Anti-obesity | Antioxidants | | | |
| Antioxidant | Antimicrobial | | | |
| Antimicrobial | Heart protection | | | |
| Heart protection | | | | |

saponins, tannins, minerals and vitamins (Bosi et al. 2019; Güçlü-Üstündağ and Mazza 2007; Kapoor 2015; Xue et al. 2016). Vitamin C is an important antioxidant, responsible for maintaining bioactive compounds like vitamin E, flavonoids and phenolics in active state. Vitamins A, C and E are involved in various body function, preventing oxidative damage of cell membranes. Legumes supply an adequate amount of minerals, where copper, zinc, chromium, iron are some of the essential micronutrients for humans that play an important role in human metabolism (Xue et al. 2016).

The antioxidant property involves scavenging of free radicals like reactive oxygen and reactive nitrogen species. Chronic diseases are often due to this free radical formation. To slower the scavenging of these radicals, high intake of antioxidants or enzymes is necessary. Therefore, consumption of these plant-based food products, which are high in the antioxidants, is recommended to relieve oxidative stress. Other bioactive compounds are named phenolics. Phenolics are broadly classified as phenolic acids, flavonoids, tannins. The activity of phenolic acids found in legumes is generally in the form of *trans*-ferulic acid, *trans*-*p*-coumaric acid and syringic acid. Legume grains were also found to be a good source of flavonoid compounds. The major role played by flavonoids involves stabilization of free radicals by donating some amount of free radicals to certain compounds (Kapoor 2015).

Studies reported that legumes contain a bundle of different phytochemicals like phenolics, tannins, carotenoids, saponins, phytosterols which provide a major source in the diet (Zhang et al. 2018). Generally speaking, phytochemicals are plant-derived compounds. These compounds are responsible for colour and other organoleptic properties of food and also exhibit several biological activities. Most of these phenolic compounds are present in the seed coats of legumes. They have the ability to interact with proteins and exhibit free radical scavenging activity. In addition, these phenolics in legumes make them a convenient source for the development of new functional foods. Polyphenols are a group which are considered as strong antioxidants that add the activity or functions of antioxidants, vitamins and enzymes as a defensive mechanism against the oxidative stress created by reactive oxygen species (Tsao 2010).

Inflammation is one of the basic mechanism in response to infection, burn or any other negative stimuli. Studies reported that legumes have been used from ancient times as a remedy to treat some inflammatory infections. All these evidence reported concludes that consumption of legumes is highly associated with the reduction of several diseases including cardiovascular disease and cancer and also play a decisive activity in improving the health of an individual (Ganesan and Xu 2017).

During the processes of soaking and cooking of legumes, proteins and water-soluble phytochemicals like saponins can be transferred in the processing water or lost by thermal degradation. For example, 0.3–2.4 g/100 g of solids were found in legume soaking water (Huang et al. 2018) and 3.3–5.6 g/100 g of solids were quantified in legume cooking water (Serventi et al. 2018; Stantiall et al. 2018) along with a 30–40% reduction of phenolics during cooking (Xu and Chang 2008).

The main objectives of this chapter is to discuss the profile and health benefits of bioactives found in legumes (phytochemicals, vitamins, peptides and lipids). This information is valuable to estimate which bioactives can be found in legume wastewater, contributing to its health benefits and potential as functional ingredient.

## 10.2 Nutrients

### *10.2.1 Phenolics*

Wastewater of legumes were shown to contain relevant quantities of phenolic compounds, ranging from 0.1 to 1.6 mg/g (Table 10.1). Phenolic compounds are among the most common secondary metabolites present in the plant kingdom. Structurally, one or more hydroxyl groups are attached to a series of aromatic rings with different side chains (Vermerris and Nicholson 2007). They are divided into subgroups of phenolic acids, flavonoids, tannins and saponins. These compounds in legumes are extensively distributed in the seed coat (flavonoids) and in the cotyledons. Highly pigmented legumes tend to have high phenols compared to the less pigmented. Phenols from legumes generally affect sensory properties and are responsible for the colour and other activities (Saucedo-Pompa et al. 2018). The level of phenolics was found to increase during germination of seeds and lowered during the cooking process (Khang et al. 2016). The total phenolic content of legumes decrease by 30–80% after soaking (Huang et al. 2018). This gradual loss of phenolics in peas and chickpeas lead to the increase in hydration rate. This phenomenon was due to the distribution and content of phenolics in the seed coat and cotyledon of the legumes. However, in some legumes, the longer soaking time made the cotyledon leach phenolics into the water. Phenolics in cooked legumes are very limited, with 30–40% of them lost with cooking (Damian et al. 2018; Xu and Chang 2008) (Table 10.2).

**Table 10.2** Total phenolic content of legume wastewater

| | Total phenolic content (mg/g) | | | |
|---|---|---|---|---|
| Legume | Soaking water | References | Cooking water | References |
| Beans | 0.3 | Huang et al. (2018) | 0.3 | Damian et al. (2018) |
| Chickpeas | 0.1–0.2 | Huang et al. (2018) and Xu and Chang (2008) | 0.6–0.7 | Damian et al. (2018) and Xu and Chang (2008) |
| Lentils | 0.4–1.6 | Huang et al. (2018) and Xu and Chang (2008) | 0.6 | Damian et al. (2018) |
| Yellow peas | 0.1–0.3 | Huang et al. (2018) and Xu and Chang (2008) | 0.6 | Damian et al. (2018) and Xu and Chang (2008) |
| Yellow soybeans | 0.1 | Huang et al. (2018) | | |

#### 10.2.1.1 Phenolic Acids

These are a type of non-flavonoid phenolic compounds which are further divided into hydroxybenzoic and hydroxycinnamic derivatives based on their ring structure. The hydroxybenzoic acid consists of C6–C1 ring structure, including gallic, vanillic, protocatechuic, p-hydroxybenzoic and syringic acid. The seed coat (hull) basically contains gallic, vanillic and syringic acids. In addition, protocatechuic and gallic are the most common ones found in almost any of the legume seeds, whereas, hydroxycinnamic acids are the aromatic compounds having a C6-C3 structure which includes caffeic, p-coumaric, trans-ferulic, sinapic and chlorogenic acid. Although, these vary greatly and their activity is seen in different legumes like beans, peas and lentils. A recent study reported that ferulic and p-coumaric are the two main types which are highly present in the legume seeds (Singh et al. 2017). Studies performed on chickpeas demonstrate that the hull isolated from coloured chickpea contained a large amount of total phenols and flavonoids which exhibited high levels of antioxidant activity. Similarly, the hulls with light colour had low level of phenolics. Hence, this variation in seed coat colour and its antioxidant activity made legumes a potential model for functional foods (Segev et al. 2011).

#### 10.2.1.2 Flavonoids

These are the main phenolics of legumes and are named as one of the largest group contributing for phenolics and also responsible for half of the known phenolic compounds. Flavonoids are the low molecular weight compounds which consist of C6–C3–C6 backbone structure with two aromatic rings. Based on the variation in the chromane ring structure are divided into subgroup as anthocyanins, flavones, flavanols, flavanones and flavans which contribute to be the main flavonoid group present in the legume seeds (Jukanti et al. 2012).

#### 10.2.1.3 Anthocyanins

Anthocyanidins and their aglycons are the main structural element of anthocyanins that majorly constitutes of aromatic rings A, B and C. When anthocyanidins are bond to a sugar moiety via glycosidic bond they are called anthocyanins. Anatomically, these anthocyanins are highly present in the coloured seed coat of the legumes. For example, the percentual amount of anthocyanins in green lentils is three to four-fold higher in the hull than in the whole seeds. Research studies showed different types of anthocyanins present in the seed coat of different legume cultivars in different countries (Singh et al. 2017). The colour of anthocyanins responsible for seed coat is pH dependent, which shows red in acidic and blue in basic conditions. However, they were found to be chemically stable in acidic conditions (Tsao 2010). In addition, antioxidant activities are expressed by anthocyanins as well prevention of diabetes (Singh et al. 2017).

#### 10.2.1.4 Flavonols and Flavones

This subgroup of flavonoids is widely distributed in the leguminous seeds. Flavonols and flavones with their three-ring structure represent the largest group of polyphenols. Flavonols (quercetin and kaempferol) and flavones (apigenin and luteolin) are the types identified under legume seeds and have different glycosidic combinations. Approximately 17–20% of the identified phenols in raw legumes are flavonols and dihydroflavonols. Kaempferol glycosides were the main types quantified in raw lentil seeds. In addition, quercetin and its derivatives are the only flavonols which are present in the coloured legume seeds and where absent in white beans. Many of the identified flavonols and flavones are concentrated in the insoluble dietary fibre fractions of the beans and lentils (Singh et al. 2017).

#### 10.2.1.5 Flavanones

The different type of flavanones obtained from legume seeds is pinocembrin, naringenin, eriodicytol, hespiritin, sakuranetin. A total estimate of 5–10% flavanones were observed in legumes (Singh et al. 2017). Several health benefits are observed from flavonoids which reduced the risk of numberous cancers including breast, lung, colon and stomach. Relatively, many of the flavonoids act against cancer in many ways by interfering with specific enzymes and hormones in the body (Jooyandeh 2011).

#### 10.2.1.6 Tannins

These are the types of polyphenols having a molecular weight of 3000 D, which are found in association with alkaloids, polysaccharides and proteins. Tannins are classified into two classes: hydrolysable and non-hydrolysable tannins (condensed tannins). These are present at a high percent in the legumes and generally occur in the outer seed coat, playing an important role in the defence mechanism of seed which usually is exposed to oxidative damage. Legumes with coloured seed coat like red and black beans contain a high amount of condensed tannins. One of the types of condensed tannins is proanthocyanidins. Among proanthocyanidins, the most important group was procyanidins which played an important role in human health and is found at a higher levels in lentils (Silva et al. 2016). In addition, the seed coat of peanuts contained higher condensed tannins compared to the kernel and cotyledons.

### *10.2.2 Peptides*

Numerous peptides found in legumes exert healthful bioactivities: anticancer, anti-inflammatory, antimicrobial, antioxidant and cardiovascular protection (Finkina et al. 2008; Kamran and Reddy 2018; Luna-Vital and de Mejía 2018; Malaguti et al. 2014; Reyes-Díaz et al. 2019; Sathoff and Samac 2019). Specifically, inhibition of acetylcholine (ACE) was demonstrated for peptides from beans, chickpeas, peas and soybeans (Kamran and Reddy 2018). Bean and soybean peptides exhibited immunological properties, while beans, chickpeas and cowpeas promoted peptide-led antimicrobial activity (Kamran and Reddy 2018). Similarly, a lentil peptide named "defensin" demonstrated a wide spectrum of antimicrobial activity against bacteria and fungi (Finkina et al. 2008; Sathoff and Samac 2019). Furthermore, beans, chickpeas and soybeans peptides acted as antioxidants (Kamran and Reddy 2018). Overall, beans and soybeans promoted the widest array of activities, including anti-inflammatory (Reyes-Díaz et al. 2019). Processing enhances legumes functionality. For instance, sprouted chickpeas exhibited higher anti-inflammatory potential than cooked chickpeas, protecting the gastrointestinal tract. Results were attributed to a synergistic effect of peptides and phenolics (Milán-Noris et al. 2018). These findings highlight the potential of wastewater from legume sprouting as source of bioactives. In addition, peptides from the non-digestible fraction of common beans (*Phaseolus vulgaris*), lunasin from soybeans (*Glycine max*) and albumin and globulin from lentils (*Lens culinaris*) were shown to prevent gastrointestinal cancer. Five mechanisms were proposed: inhibition of metalloproteases, disruption of mitochondria, damage to DNA, interaction with membrane receptors and induced sensitivity to apoptosis (Luna-Vital and de Mejía 2018). The protein content of legume wastewater was shown to range from to 0.1 to 1.5 g/100 g (Buhl et al. 2019; Huang et al. 2018; Meurer et al. 2020; Serventi et al. 2018; Stantiall et al. 2018), representing up to 30% of the solid fraction. Therefore, legume wastewater represents a significant source of nutritionally relevant peptides.

**Table 10.3** Saponin content of legume wastewater

| | Saponin content (mg/g) | | | |
|---|---|---|---|---|
| Legume | Soaking water | References | Cooking water | References |
| Beans | 3.4 | Huang et al. (2018) | 7.9 | Damian et al. (2018) |
| Chickpeas | 1.0 | Huang et al. (2018) | 12 | Damian et al. (2018) |
| Lentils | 3.0 | Huang et al. (2018) | 14 | Damian et al. (2018) |
| Yellow peas | 3.2 | Huang et al. (2018) | 9.8 | Damian et al. (2018) |
| Yellow soybeans | 0.9 | Huang et al. (2018) | 6.4 | Serventi et al. (2018) |

### *10.2.3 Saponins*

Saponins are a group of compounds characterized by a carbohydrate moiety and an aglycone. Moderate amounts of saponins were detected in the soaking and cooking water of five legumes ranging from 0.9 mg/g of soybean soaking water to 14 mg/g of lentil cooking water (Table 10.3). Saponins in legumes have growing interest because of their ability to lower the plasma cholesterol level in humans and their activity to suppress cancer growth. Among all the legume types, lentils are a best-considered source for saponins, whereas, soyasaponin I and IV are of main types. A study demonstrated that saponins from soybeans have a unique structure and are found to have various biological activities (Dixit et al. 2011). New research is focusing on saponin structure and their clinical studies which have shown various health benefits. Clinical study reports that saponins have the ability to lower blood cholesterol, blood lipids and glucose level and also reduce the risk of cancer. Saponins are amphiphilic in nature and are known to be bitter in taste. Soy, chickpea, peas are a good source of saponins. Although a high percentage of saponins is present in chickpea compared to other legumes and the content was decreased by 50% after boil. Interestingly, a study showed that saponin content was associated with protein content. Hence, this finding clearly explains the formation and association of protein-saponin complexes in legumes and also determined the presence of saponins in protein fractions of legume seeds (Güçlü-Üstündağ and Mazza 2007). In addition, saponins act as surfactants which stabilize emulsions, where haricot and split yellow peas found to have high saponin content. A study demonstrated the loss of saponins from legumes during the soaking process, and also indicated that almost 50% of saponins was decreased during a 16 hour soaking of legumes (Huang et al. 2018). Similarly the percentage of saponins in haricot beans, green lentils and split yellow peas where three times more than other legumes (Damian et al. 2018).

### *10.2.4 Vitamins*

Legumes contribute as a good source of vitamins. They are required in a small amount in the diet and are met with the daily diet from legumes, pulses. Among all the legumes present, chickpea contributes for the good source of vitamins like folic acid and tocopherols. It is one of the good sources of folic acid which is coupled with higher amounts of water soluble vitamins like riboflavin (B2), pantothenic acid (B5) and pyridoxine (B6). Research done on vitamins confirmed that soybeans are the richest source of vitamin B (Dixit et al. 2011).

The seeds of legumes contain carotenoids, tocopherols, and fatty acids that supply nutrients for humans, comprising vitamin precursors. The major finding showed that soybean and chickpea are rich in β-carotene. Carotenoids are a type of natural pigment precursors of vitamin A that play an important role in the reduction of various diseases in humans. In addition, tocopherols are also a group of a compound which has vitamin E activity. Although, from all the forms of tocopherol, γ-isoform is the most commonly found in the seed coat of legumes like lentils, chickpeas and broad beans. Tocopherols play an important role in the anti-inflammatory process and deficiency of this vitamin results in a wide range of disorders. Several epidemiological studies suggested that the activity of this vitamin may protect against cardiovascular diseases (Silva et al. 2016). Current developments suggest all these types of tocopherols as emerged in the form of vitamin E molecules with major functions in health and diseases. Even at nanomolar concentration, the α-tocotrienol have the ability to prevent neurodegeneration (Colombo 2010). To the best of our knowledge, vitamin content of legume wastewater has not been investigated by any study, therefore future research on this nutrient group is warranted.

## *10.2.5 Lipids*

Apart from minerals, vitamins, and phenolic content, legumes are also a source of lipids. The lipid fraction of seeds determines most of their energy and nutritional content. There are only limited studies conducted on the lipid profile or fatty acid profile of legumes. A study confirmed the palmitic acid, oleic acid, stearic acid were heterogenous among all the legumes. Chickpeas contained a high amount of oleic acid. In contrast, polyunsaturated fatty acids (PUFA) were present in high amount in beans and lentils, which had several beneficial effects on cardiovascular diseases including blood lipid profile, improved insulin sensitivity and showed a lower incidence of type 2 diabetes. Linoleic acid was the most important type of PUFA present in chickpea, soybean and fava beans (Caprioli et al. 2016).

### 10.2.5.1 Phospholipids and Phytosterols

The amount of phospholipids in legumes is 1–3% of the total lipids. Whereas, phytosterols are the natural components of the plant cell membranes. A higher level of phytosterols is present in the outer kernel or seed coat of the legumes. They are primarily β-sitosterol, campesterol and stigmasterol. Legumes including lentils are one of the important sources of phytosterols which meet the daily dietary requirements. Sitosterol was found to be one of the most prevalent phytosterols in lentils.

They had shown to have various biological activities which include anti-inflammatory, antioxidative, anticarcinogenic activities. Furthermore, studies confirmed the inhibitory activity of phytosterols towards the intestinal absorption of cholesterol, thus lowering the concentration of plasma cholesterol, specifically LDL (low density lipoprotein) cholesterol (Zhang et al. 2018). Nonetheless, the current research on legume wastewater has shown no traces of lipids (Huang et al. 2018; Serventi et al. 2018; Stantiall et al. 2018).

## 10.3 Biological Activities

### *10.3.1 Antioxidant Property*

Reactive oxygen species (ROS) such as free radicles and peroxides, are naturally occurring by-products by different metabolic processes. These may lead to cellular damage and result in the development of the risk of several diseases. Several studies have shown that certain diets reduce the risk of diseases related to oxidative stress. Compared to other legumes, chickpea is considered to be a source with high antioxidant property (Torres-Fuentes et al. 2015). However, the antioxidant activity depends upon the protein source and amino acid composition. Studies showed that antioxidant peptides exert metal chelation activity which allows them to interact with certain free radicles, thus preventing radical formation, reducing oxidative stress and risk of various diseases like cancer (Kou et al. 2013). Several studies quantified the amount of antioxidants in legumes before and after cooking. The radical scavenging activity of seed coats was affected by the cooking time (Rocha-Guzmán et al. 2007). Generally, phenolic compounds have the ability to donate hydrogen atoms and scavenge free radicles. The antioxidant activity exhibited by legumes like fava beans, red beans, broad bean and red lentil green lentils was reported to be high. Chickpeas with black and red coloured seed coat showed antioxidant potential when compared to chickpeas with light coloured. In addition, germination can further increase the activity of antioxidant in legumes due to the presence of phenolic compounds. Consequently, the presence of high antioxidant potential in legumes was useful in the production of healthy foods (Singh et al. 2017).

### *10.3.2 Anti-Inflammatory Activity*

Inflammation is the basic defence mechanism of response to infection, burn and other diseases. Cyclooxygenase (COX) and lipoxygenase (LOX) are the two enzymes mediating the mechanism of inflammation. Enzymes COX-1 and COX-2 are the two types regulating the inflammatory response. COX-2 increases the risk of inflammation, whereas COX-1 decreases it. In addition, COX-2 is involved in the production of prostaglandins in response to stimuli (Kim et al. 2017). However, legumes have been since used by ancient treatment remedies as a potential source to treat some inflammatory symptoms like skin infections by legume water paste and also a treatment for burns. Similarly, regular consumption of legume foods, particularly chickpeas, lentils and soybeans have been proven to reduce the risk of developing chronic inflammatory diseases including cancer (Papandreou et al. 2019). Phenol rich legumes like lentils have the potential to reduce blood pressure because of its activity of angiotensin converting enzyme. Studies also suggested that the hull of the common beans exhibited high inflammatory activity by inhibiting the activity of LOX enzyme (Oomah et al. 2010). Specifically, plant-derived phenolics and flavonoids exhibited good anti-inflammatory activity by regulating the level of certain cytokines and interleukins. Hence, both the antioxidant and anti-inflammatory activities are positively associated with phenolic content (Zhang et al. 2018).

### 10.3.3 Anticancer Activity

Consumption of legumes may reduce the risk of cancer (de Cedrón et al. 2018; Papandreou et al. 2019). The inhibition of cancer cells by commonly consumed legume foods have been researched and studied by many groups. Interestingly, various types of legumes like chickpea hydrolysates functioned in the inhibition of human epithelial adenocarcinoma cells and leukaemia cells (Guardado-Félix et al. 2019; Rao et al. 2018). The processes of inflammation and cancer are connected, where chronic inflammation leads individuals to the development of several types of cancer. Studies also suggested that suppression of inflammatory pathways may provide alternate ways for the prevention of cancer by inhibiting the activity of macrophages (López-Barrios et al. 2014). The high fibre content of legumes was associated with prevention of cancer because of its anti-proliferative activity and induction of apoptosis of cells which reduces the risk of cancer (Sánchez-Chino et al. 2015; Zhu et al. 2015). The anticancer activity of legumes is also correlated to their mineral content (zinc, selenium, calcium and others) which decreases the oxidative stress and inhibits the development of tumour cells (Ohigashi et al. 2013). The biologically active components of legumes tannins, phytic acid and saponins are predominantly involved in anticancer activity. Specifically, saponins have found to have a major role in the suppression of tumour cells by inhibiting the activity of carcinoma cells (Lima et al. 2016). Regarding tannins and phytic acid, the anticancer activity of these phytochemicals may be related to their phytoestrogen function (Barman et al. 2018; Slavin et al. 1999). Mechanistically, some of the phytochemicals present in soybeans or chickpea exhibit their effect by binding to human estrogenic receptors. Numerous hormone-related cancers develop as a result of a hormonal imbalance between estrogen and androgen, which may increase the risk of cancer. Isoflavones exhibited similar function to these hormones, having the ability to bind to human estrogen receptor by exhibiting powerful effects on hormone-dependent tumours (Kapoor 2015).

### 10.3.4 Cardio Protective Effect of Legumes

Evidence from numerous studies has shown that consumption of legumes is associated with reduction of cardiovascular diseases. Several factors were linked to cardiovascular diseases, for example lipid profile, blood pressure and inflammation. Consumption of legumes has resulted in lowering the total cholesterol level and also reduce blood pressure by angiotensin converting enzyme. Besides that, phenol-rich legumes like lentils, soybeans, and chickpeas express properties of anti-hyperlipidaemia, anti-cholesterolemia and cardio protective, thus reducing the risk of coronary heart diseases and hypertension. Consequently, consumption of legumes was recommended on daily basis in order to reduce the risk of cardiovascular diseases (Ganesan and Xu 2017).

The high fibre content and low glycemic index of legumes were involved in lowering the level of LDL-cholesterol (Low Density Lipoprotein). Isoflavones from legumes stimulate the secretion of adiponectin, a hormone which exhibits the property of anti-inflammatory property of blood vessels and is linked with the reduction of a heart attack when it is in high level (Kapoor 2015).

### 10.3.5 Antimicrobial Activity

Legumes containing flavonoids and other phytochemicals were regarded as safe for use in diagnostic kits. Studies reported that a major bioactive peptide called "Defensin" isolated from legume seeds, especially lentils, have wide spectrum activity, including microbial activity against various infectious diseases caused by bacteria and fungi (Finkina et al. 2008; Sathoff and Samac 2019). Biologically, defensin is synthesized in lentil seeds and are involved in the development of innate immunity. This peptide, along with phytochemicals, have a potential to inhibit microbial growth. Defensin expresses its activity by interupting the viral digestive enzymes which results in viral replication. Hence, these have been further observed in blocking the ion channels. Therefore, the activity of defensing from the legume seeds in combination with other phytochemicals like phenolic compounds act as a major inhibitor in microbial growth (Ganesan and Xu 2017).

The hulls of legume seeds contain various potential activities and are determined as a source of natural additives. However, a limited amount of information is available regarding antimicrobial activity. Legume hull extracts have the capability to inhibit the activity of certain bacteria, where some are most effective and beneficial. In addition, this may be because of the presence of high polyphenols in the hull of the legumes, which states that polyphenols are responsible for microbial inhibition. Hence, several studies showed that gram negative bacteria are more resistant to the polyphenols present in the legumes compared to gram positive bacteria, this may be because of its different cell wall composition (Kanatt et al. 2011).

### 10.3.6 *Anti-Obesity Activity*

High intake of legumes is inversely correlated with obesity and diabetes. Researchers suggest that legumes containing flavonoids, saponins and fibre complements satiety and helps in lowering the amount of food intake which finally results in maintaining the body weight of obese patients. Legumes like lentils have the ability to inhibit α-glucosidase and pancreatic lipase enzyme, which finally results in the reduction of glucose and fat digestion. Relatively, the flavonoid content in legumes has the potential to inhibit the activity of several harmful enzymes and manage the blood glucose and body weight (Ganesan and Xu 2017).

## 10.4 Conclusions

Legumes consist of a vast array of bioactive compounds such as phenolics, peptides, saponins, vitamins and lipids. These nutrients express several properties such as anticancer, anti-inflammatory, anti-obesity, antioxidant, antimicrobial and heart protection. The seed coat of legumes contain high level of phenolic compounds that exhibit high antioxidant activities. However, industrial processes of soaking, cooking and sprouting may decrease the level of these compounds which also results in decrease of overall antioxidant activity. The great loss of phytochemicals like saponins, phenolics, minerals in soaking and cooking water varied significantly based on the type of the legume and also the structure of the seed. Phenolics, saponins and peptides were found in the processing water of legumes, while vitamins have not been investigated. Therefore, isolation of these nutrients from legume wastewater may allow recovery of bioactives expressing numerous health benefits.

**Acknowledgments** This realisation of this chapter was possible thank to the teaching funds allocated by Lincoln University to the Taught Master course named "FOOD 698 – Research Essay".

## References

Barman, A., Marak, C.M., Barman, R.M., & Sangma, C.S. (2018). Nutraceutical properties of legume seeds and their impact on human health. In *Legume seed nutraceutical research*. IntechOpen.

Bosi, S., Bregola, V., Dinelli, G., Trebbi, G., Truzzi, F., & Marotti, I. (2019). The nutraceutical value of grain legumes: Characterisation of bioactives and antinutritionals related to diabesity management. *International Journal of Food Science & Technology, 54*(10), 2863–2871.

Buhl, T. F., Christensen, C. H., & Hammershøj, M. (2019). Aquafaba as an egg white substitute in food foams and emulsions: Protein composition and functional behavior. *Food Hydrocolloids, 96*, 354–364.

Caprioli, G., Giusti, F., Ballini, R., Sagratini, G., Vila-Donat, P., Vittori, S., & Fiorini, D. (2016). Lipid nutritional value of legumes: Evaluation of different extraction methods and determination of fatty acid composition. *Food Chemistry, 192*, 965–971.

Colombo, M. L. (2010). An Update on Vitamin E, Tocopherol and Tocotrienol—Perspectives. *Molecules, 15*(4):2103–2113.

Damian, J. J., Huo, S., & Serventi, L. (2018). Phytochemical content and emulsifying ability of pulses cooking water. *European Food Research and Technology, 244*(9), 1647–1655.

de Cedrón, M. G., de Molina, A. R., & Reglero, G. (2018). Legumes and cancer. In *Legumes* (pp. 324–349), Royal Society of Chemistry, Croydon, UK.

Dixit, A.K., Antony, J., Sharma, N.K., & Tiwari, R.K. (2011). 12. Soybean constituents and their functional benefits. *Research Singpost, 37*(2), 661.

Finkina, E. I., Shramova, E. I., Tagaev, A. A., & Ovchinnikova, T. V. (2008). A novel defensin from the lentil Lens culinaris seeds. *Biochemical and Biophysical Research Communications, 371*(4), 860–865.

Ganesan, K., & Xu, B. (2017). Polyphenol-rich lentils and their health promoting effects. *International Journal of Molecular Sciences, 18*(11), 2390.

Guardado-Félix, D., Antunes-Ricardo, M., Rocha-Pizaña, M. R., Martínez-Torres, A. C., Gutiérrez-Uribe, J. A., & Saldivar, S. O. S. (2019). Chickpea (Cicer arietinum L.) sprouts containing supranutritional levels of selenium decrease tumor growth of colon cancer cells xenografted in immune-suppressed mice. *Journal of Functional Foods, 53*, 76–84.

Güçlü-Üstündağ, Ö., & Mazza, G. (2007). Saponins: Properties, applications and processing. *Critical Reviews in Food Science and Nutrition, 47*(3), 231–258.

Huang, S., Liu, Y., Zhang, W., Dale, K. J., Liu, S., Zhu, J., & Serventi, L. (2018). Composition of legume soaking water and emulsifying properties in gluten-free bread. *Food Science and Technology International, 24*(3), 232–241.

Jooyandeh, H. (2011). Soy products as healthy and functional foods. *Middle-East Journal of Scientific Research, 7*(1), 71–80.

Jukanti, A. K., Gaur, P. M., Gowda, C., & Chibbar, R. N. (2012). Nutritional quality and health benefits of chickpea (*Cicer arietinum L.*): A review. *British Journal of Nutrition, 108*(S1), S11–S26.

Kamran, F., & Reddy, N. (2018). Bioactive peptides from legumes: Functional and nutraceutical potential. *Recent Advances in Food Science, 1*(3), 134–149.

Kanatt, S. R., Arjun, K., & Sharma, A. (2011). Antioxidant and antimicrobial activity of legume hulls. *Food Research International, 44*(10), 3182–3187.

Kapoor, S. (2015). Bioactives and therapeutic potential of legumes: A review. *International Journal of Pharmacy and Biological Sciences, 5*, 65–74.

Khang, D., Dung, T., Elzaawely, A., & Xuan, T. (2016). Phenolic profiles and antioxidant activity of germinated legumes. *Food, 5*(2), 27.

Kim, M. J., Shrestha, S., Eldridge, M., Cortes, M., Singh, P., Liow, J. S., Gladding, R., Zoghbi, S., Fujita, M., Pike, V., & Innis, R. (2017). Novel PET radioligands show that, in rhesus monkeys, COX-1 is constitutively expressed and COX-2 is induced by inflammation. *Journal of Nuclear Medicine, 58*(1), 203–203.

Kou, X., Gao, J., Xue, Z., Zhang, Z., Wang, H., & Wang, X. (2013). Purification and identification of antioxidant peptides from chickpea (*Cicer arietinum L.*) albumin hydrolysates. *LWT-Food Science and Technology, 50*(2), 591–598.

Lima, A. I. G., Mota, J., Monteiro, S. A. V. S., & Ferreira, R. M. S. B. (2016). Legume seeds and colorectal cancer revisited: Protease inhibitors reduce MMP-9 activity and colon cancer cell migration. *Food Chemistry, 197*, 30–38.

López-Barrios, L., Gutiérrez-Uribe, J. A., & Serna-Saldívar, S. O. (2014). Bioactive peptides and hydrolysates from pulses and their potential use as functional ingredients. *Journal of Food Science, 79*(3), R273–R283.

Luna-Vital, D., & de Mejía, E. G. (2018). Peptides from legumes with antigastrointestinal cancer potential: Current evidence for their molecular mechanisms. *Current Opinion in Food Science, 20*, 13–18.

Malaguti, M., Dinelli, G., Leoncini, E., Bregola, V., Bosi, S., Cicero, A. F., & Hrelia, S. (2014). Bioactive peptides in cereals and legumes: Agronomical, biochemical and clinical aspects. *International Journal of Molecular Sciences, 15*(11), 21120–21135.

Margier, M., Georgé, S., Hafnaoui, N., Remond, D., Nowicki, M., Du Chaffaut, L., ... & Reboul, E. (2018). Nutritional composition and bioactive content of legumes: Characterization of pulses frequently consumed in France and effect of the cooking method. *Nutrients, 10*(11), 1668.

Meurer, M. C., de Souza, D., & Marczak, L. D. F. (2020). Effects of ultrasound on technological properties of chickpea cooking water (aquafaba). *Journal of Food Engineering, 265*, 109688.

Milán-Noris, A. K., Gutiérrez-Uribe, J. A., Santacruz, A., Serna-Saldívar, S. O., & Martínez-Villaluenga, C. (2018). Peptides and isoflavones in gastrointestinal digests contribute to the anti-inflammatory potential of cooked or germinated desi and kabuli chickpea (*Cicer arietinum L.*). *Food Chemistry, 268*, 66–76.

Moughan, P. J. (2009). Digestion and absorption of proteins and peptides. In *Designing functional foods* (pp. 148–170). Woodhead Publishing, Cambridge, UK.

Ohigashi, H., Osawa, T., Terao, J., Watanabe, S., & Yoshikawa, T. (Eds.). (2013). *Food factors for cancer prevention*. Springer Science & Business Media, Tokyo, Japan.

Oomah, B. D., Corbé, A., & Balasubramanian, P. (2010). Antioxidant and anti-inflammatory activities of bean (*Phaseolus vulgaris L.*) hulls. *Journal of Agricultural and Food Chemistry, 58*(14), 8225–8230.

Papandreou, C., Becerra-Tomás, N., Bulló, M., Martínez-González, M. Á., Corella, D., Estruch, R., Ros, E., Arós, F., Schroder, H., Fitó, M., & Serra-Majem, L. (2019). Legume consumption and risk of all-cause, cardiovascular, and cancer mortality in the predimed study. *Clinical Nutrition, 38*(1), 348–356.

Rao, S., Chinkwo, K., Santhakumar, A., & Blanchard, C. (2018). Inhibitory effects of pulse bioactive compounds on cancer development pathways. *Diseases, 6*(3), 72.

Reyes-Díaz, A., Del-Toro-Sánchez, C. L., Rodríguez-Figueroa, J. C., Valdéz-Hurtado, S., Wong-Corral, F. J., Borboa-Flores, J., González-Osuna, M. F., Perez-Perez, L. M., & González-Vega, R. I. (2019). Legume proteins as a promising source of anti-inflammatory peptides. *Current Protein & Peptide Science, 20*(12), 1204–1217.

Rocha-Guzmán, N. E., González-Laredo, R. F., Ibarra-Pérez, F. J., Nava-Berumen, C. A., & Gallegos-Infante, J.-A. (2007). Effect of pressure cooking on the antioxidant activity of extracts from three common bean (*Phaseolus vulgaris L.*) cultivars. *Food Chemistry, 100*(1), 31–35.

Roy, F., Boye, J., & Simpson, B. (2010). Bioactive proteins and peptides in pulse crops: Pea, chickpea and lentil. *Food Research International, 43*(2), 432–442.

Sánchez-Chino, X., Jiménez-Martínez, C., Dávila-Ortiz, G., Álvarez-González, I., & Madrigal-Bujaidar, E. (2015). Nutrient and nonnutrient components of legumes, and its chemopreventive activity: A review. *Nutrition and Cancer, 67*(3), 401–410.

Sathoff, A. E., & Samac, D. A. (2019). Antibacterial activity of plant defensins. *Molecular Plant-Microbe Interactions, 32*(5), 507–514.

Saucedo-Pompa, S., Martínez-Ávila, G. C. G., Rojas-Molina, R., & Sánchez-Alejo, E. J. (2018). Natural beverages and sensory quality based on phenolic contents. In *Antioxidants in foods and its applications* (p. 69), IntechOpne, Croatia.

Segev, A., Badani, H., Galili, L., Hovav, R., Kapulnik, Y., Shomer, I., & Galili, S. (2011). Total phenolic content and antioxidant activity of chickpea (*Cicer arietinum L.*) as affected by soaking and cooking conditions. *Food and Nutrition Sciences, 2*(07), 724.

Serventi, L., Wang, S., Zhu, J., Liu, S., & Fei, F. (2018). Cooking water of yellow soybeans as emulsifier in gluten-free crackers. *European Food Research and Technology, 244*(12), 2141–2148.

Silva, L. R., Peix, A., Albuquerque, C., & Velàzquez, E. (2016). *Bioactive compounds of legumes as health promoters. Natural bioactive compounds from fruits and vegetables* (pp. 3–26). Sharjah: Bentham Science Publishers.

Singh, B. P., Vij, S., & Hati, S. (2014). Functional significance of bioactive peptides derived from soybean. *Peptides, 54*, 171–179.

Singh, B., Singh, J. P., Kaur, A., & Singh, N. (2017). Phenolic composition and antioxidant potential of grain legume seeds: A review. *Food Research International, 101*, 1–16.

Slavin, J. L., Martini, M. C., Jacobs, D. R., Jr., & Marquart, L. (1999). Plausible mechanisms for the protectiveness of whole grains. *The American Journal of Clinical Nutrition, 70*(3), 459s–463s.

Stantiall, S. E., Dale, K. J., Calizo, F. S., & Serventi, L. (2018). Application of pulses cooking water as functional ingredients: The foaming and gelling abilities. *European Food Research and Technology, 244*(1), 97–104.

Torres-Fuentes, C., del Mar Contreras, M., Recio, I., Alaiz, M., & Vioque, J. (2015). Identification and characterization of antioxidant peptides from chickpea protein hydrolysates. *Food Chemistry, 180*, 194–202.

Tsao, R. (2010). Chemistry and biochemistry of dietary polyphenols. *Nutrients, 2*(12), 1231–1246.

Vermerris, W., & Nicholson, R. (2007). *Phenolic compound biochemistry*. Springer Science & Business Media.

Xu, B., & Chang, S. K. (2008). Effect of soaking, boiling, and steaming on total phenolic content and antioxidant activities of cool season food legumes. *Food Chemistry, 110*(1), 1–13.

Xue, Z., Wang, C., Zhai, L., Yu, W., Chang, H., Kou, X., & Zhou, F. (2016). Bioactive compounds and antioxidant activity of mung bean (*Vigna radiata L.*), soybean (*Glycine max L.*) and black bean (*Phaseolus vulgaris L.*) during the germination process. *Czech Journal of Food Science, 34*(1), 68-78.

Zhang, B., Peng, H., Deng, Z., & Tsao, R. (2018). Phytochemicals of lentil (Lens culinaris) and their antioxidant and anti-inflammatory effects. *Journal of Food Bioactives, 1*, 93–103. -193–103.

Zhu, B., Sun, Y., Qi, L., Zhong, R., & Miao, X. (2015). Dietary legume consumption reduces risk of colorectal cancer: Evidence from a meta-analysis of cohort studies. *Scientific Reports, 5*, 8797.

# Chapter 11
# Edible Packaging from Legume By-Products

**Yanyu Zhang and Luca Serventi**

## 11.1 Introduction

Biodegradable plastic is considered a replacement for traditional material in packaging, with increasing demand in these years. Circular economy of legumes is a promising approach to environmentally friendly food processing. Bioplastics are viewed as a sustainable material in terms of the global trend of using eco-friendly materials in food packaging (Korhonen et al. 2018). Since traditional plastic needs hundreds of years to be degraded and decay in the soil, it places a heavy burden on the environment (Yadav et al. 2018). According to the result of research, in 2015 the global plastic industry has produced approximately 322 million tons of waste, increasing by 3.5% from 2014 (European Bioplastics 2016). The possibility of replacement of synthetic conventional packaging by soluble packaging material has been proven in a recent study (Puscaselu et al. 2019). Bioplastics are important in tackling environmental challenges caused by traditional plastics. Not only can it reduce dependence on fossil fuels and associated environmental influence, but it can also help overcome the waste disposal obstacle (Stahel 2016). Bioplastics are made from natural polymers obtained entirely from renewable biomass that can be reduced by the natural processes of microorganisms (Geissdoerfer et al. 2017). Therefore, bioplastics can be made from inexpensive, readily available materials, covering derived starch or spontaneously occurring polymeric cellulose (Orenia et al. 2018). In addition, bioplastics can be used in many fields such as agriculture and medicine. Further research into the possibility of using more materials can help reduce recycling costs and will compete with petrochemicals in the future (Thakur et al. 2018). Comparing with traditional materials, bioplastics such as PLA

Y. Zhang · L. Serventi (✉)
Department of Wine, Food and Molecular Biosciences, Faculty of Agriculture and Life Sciences, Lincoln University, Lincoln, Christchurch, New Zealand
e-mail: Luca.Serventi@lincoln.ac.nz



L. Serventi, *Upcycling Legume Water: from wastewater to food ingredients*,
https://doi.org/10.1007/978-3-030-42468-8_11

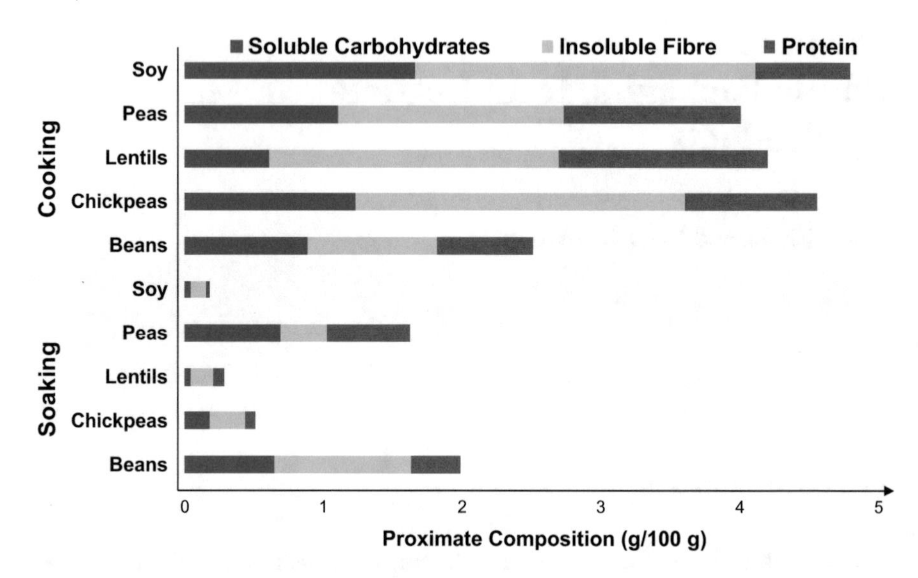


**Fig. 11.1** Fibre and protein content of soaking and cooking water of five different legumes (Huang et al. 2018; Serventi et al. 2018; Stantiall et al. 2018)

(polylactide), starch, PHA (polyhydroxyalkanoates) and cellulose can be applied to packaging for both short and long shelf-life food products (Peelman et al. 2013).

Legumes are rich in vegetable protein (Boye et al. 2010) and fibre (Tosh and Yada 2010). Various types of plant proteins such as soybeans, corn and others are used in the creation of bioplastics. In addition, legumes contain numerous bioactive compounds with surfactant properties such as phenolics and saponins (Sreerama et al. 2010). According to current research, protein-based biomaterials are one of the rapidly degrading polymers. Because of its unique structure and wide range of functional properties, the use of protein to make bioplastics could be a research worthy program in the future (Jerez et al. 2007). In addition, dietary fibre consists of insoluble fibre, soluble fibre, indigestible oligosaccharide, and resistant starch (Venkidasamy et al. 2019). All legume fibres consist of cellulose and different fibres have different components and structures, depending on legume types, growing situation and various soil position. Nowadays, soy-based biocomposites strengthened with both soluble and insoluble fibre were suggested to be a higher value source than PP synthesis one (Gurunathan et al. 2015). The processing water of legumes (wastewater from soaking and cooking) were shown to contain relevant amounts of soluble fibre, insoluble fibre and protein (Fig. 11.1).

In what follows, it was hypothesized that legume wastewater could be used as source of bioplastics and possibly edible packaging. Fibre and protein have been found in legume wastewater from soaking, boiling and canning. Therefore, these by-products could be a source of raw material for the manufacturing of biodegradable plastics. The aim of this book chapter is to describe technologies for the production of edible, biodegradable packaging from legume fibre, protein and phytochemicals.

## 11.2 Bioplastic from Fibre

### 11.2.1 Isolation of Fibre from Legumes

In industrial production, where plastics can be used as food packaging, the purpose of separating the legume fibres is usually achieved by a technique called fractionation (Martens et al. 2017). When fibres similar to those found in the legumes soaking water are fractionated under high temperature conditions, the separated insoluble matter is removed from the water through a sieve (Table 11.1). In addition, the insoluble fibre concentrate separated at the top of the fractionation column is further subjected to freeze-drying sieving, and the insoluble fibre concentrate is purified by digesting the remaining starch (Kutoš et al. 2003). The purification process mainly consists of dispersing the insoluble fibre concentrate in water, shaking intermittently

**Table 11.1** Recent technologies used to extract and manufacture legume fibre and protein into bioplastics

| Nutrient | Extraction | References | Manufacturing | References |
|---|---|---|---|---|
| Fibre | ***Sieves*** | Dalgetty and Baik (2003) | ***Compression molding*** | Deshpande et al. (2000) |
| | ***Fractionation columns*** | Martens et al. (2017) | ***Microbial transglutaminase*** (MTGase) | Gaspar and de Góes-Favoni (2015) |
| | ***Freeze-drying sieving*** | Kutoš et al. (2003) | | |
| Protein | ***Centrifugation*** | Megías et al. (2016) | ***Plasticizers*** (polyols, fatty acids, mono-, di- and oligosaccharides, citrate esters) | Suderman et al. (2018) |
| | ***Filtration/ Ultrafiltration*** | Cho and Rhee (2004) | ***Surfactants*** (SDS) | Wu et al. (2018) |
| | | | ***Biodegradable Polymers*** (caseinate, chitin, propylene glycol alginate) | Fabra et al. (2010), Li and Pelton (2005), and Zheng et al. (2003) |
| | | | ***Oils*** (cinnamon oil, ginger oil) | Atarés et al. (2010), Hasheminya et al. (2019), and Kim et al. (2015) |
| | | | ***Chemical cross-linking*** (calcium sulfate and chloride) | Park et al. (2001), Sabbah et al. (2019), Vaz et al. (2003) |
| | | | ***Radiation modification*** (UV) | Boy et al. (2018) |
| | | | ***Enzyme cross-linking*** (horseradic peroxidis) | Mohammad Zadeh et al. (2018) |
| | | | ***Surface modification*** (PLA) | Rhim et al. (2007) and Zhang et al. (2016) |

in a 100 °C water bath, treating it with heat-stable α-amylase for 30 minutes at pH 6 and finally separating it by centrifugation, then the completed insoluble fibre concentrate is purified (Dalgetty and Baik 2003).

Edible packaging from legumes cellulose is a novel biodegradable plastic type in food packaging. The technology of extracting cellulose from beans as a raw material for plastics is now improving. When temperature reaches about 140 °C at 20Mpa of pressure for 7 minutes, the structure of plastic is fully developed (Salmoral et al. 2000). In this study, legumes were soaked in 0.005 M NaHSO3, homogenized and crushed to obtain the supernatant for the subsequent purification process. After cellulose oxidation and cellulose-starch bioplastic preparation by solution casting and evaporation process, the crystallinity of cellulose could be detected (Cifriadi et al. 2017). As a result, the bioplastic compound made from cellulose formed an excellent edible plastic material for food packaging. When blending soy protein isolation with carboxymethylcellulose (CMC), the bioplastic film showed enhanced tensile strength and water solubility (Han et al. 2015).

### *11.2.2 Application of Bioplastic from Fibre*

Two main nutrients found in legumes, fibre and protein, are used to manufacture bioplastics through different mechanisms (Table 11.1). The following sections discuss the scientific and technological details of these technologies. Nowadays, natural fibres are mainly used in various fields as functional, low-cost and renewable materials. Bioplastic covers three main types of degradable polymeric materials, polysaccharides, proteins and PHA, which are all from renewable sources. Based on the compostability of the fibres in legumes, they can be a practical explanation to the waste treatment problem of polymeric substance. As a renewable resource, bioplastic material is also of great positive significance to environment on earth (Gurunathan et al. 2015). For instance, the extraction of bamboo fibres for biodegradable plastics has been studied and put into use widely. The extraction technology consists on conventional methods of compression molding technique (CMT) and it tends to mature in recent years (Deshpande et al. 2000). These plants with high strengths fibres are heavy load durable and biodegradable. The eco-friendly packaging made from bamboo is a worldwide prevailing material in food packaging. Bioplastic extracted from fibre-rich plants has advanced undoubtedly over the past few years. They take the advantages of eco-friendly, biodegradability, high strength, low cost and sustainability. In terms of the whole development, bioplastic materials with the reinforcement of insoluble fibre cover different kinds of matrices used for the composites, an accomplishment of biodegradable materials in practical industry and drives distinctive fabrication methods. Biocomposites extracted from bamboo fibres have been widely used in various facility and tools like biodegradable bowls, stews and the bamboo charcoal which has the advantage to burn without smoke and fumes. The major constituents of bamboo are cellulose, hemicellulose and lignin. Lignin and cellulose form a stable structure in fibre, which gives the bioplastic its stability and plasticity (Nurul Fazita et al. 2016).

Although the amount and type of insoluble carbohydrates vary from different kinds of legumes, insoluble fibre in both soaking and cooking water is the major solid fraction. Comparing with the containing fibre distribution in bamboo, legumes contain both soluble fibre and insoluble fibres. Generally, the soluble fibres including pectin, gums, mucilages account for a small percentage (Oomah et al. 2011). On the contrary, insoluble dietary fibre accounts for more than 50% in the composition of legumes like cellulose, lignin and some hemicellulose (Oomah et al. 2011). In that case, legumes may contribute not only to the nutrient food supply industry but also in producing of environment-friendly materials for packaging, utensil, the facility even fuel product due to the large percentage of dietary fibre proportion. According to the study (Dalgetty and Baik 2003), the total dietary fibre contents of pea hulls ranged from 89.9% to 91.6% and the dominant discrepancy between cotyledon fibres and hull fibre is that the main component of hulls are cellulose, hemicelluloses and lignin, while for cotyledon fibres, they generally consist non-structural polysaccharides such as gums and pectins. As a result, insoluble fibres from husk commit a crucial performance in acting as the raw material of bioplastic instead of cotyledon fibres in legumes (Brummer et al. 2015).

### *11.2.3 Acceptability of Bioplastic from Legumes Fibre*

Cellulosic bioplastics primarily consist of the cellulose esters and their derivatives, including celluloid. Significant amounts of fibre, soluble and insoluble, were found in the soaking and cooking water of various legumes (Huang et al. 2018; Serventi et al. 2018; Stantiall et al. 2018). Therefore, high feasibility occurs for cellulose to form the biocomposites by degrading to give glucose the repeating and smallest unit with a chemi-mechanical technique to isolate the cellulose from legumes (Sood and Dwivedi 2017). The amount of cellulose determines the mechanical properties of biomaterials (Mukherjee and Kao 2011). According to a previous experiment, the cellulose and fibre-reinforced cellulose obtained from pea flour were improved, possessing higher value than the bioplastic based on corn starch mixed with vegetal oils. In that case, it could be concluded that the bioplastics extracted from legumes fibre are more mechanically resistant and more rigid than that from corn fibres because of the presence of microbial transglutaminase (Gaspar and de Góes-Favoni 2015), which is active over 40 °C and pH between 7 and 7.5 (Giosafatto et al. 2018).

## 11.3 Bioplastic from Protein

### *11.3.1 Isolation of Protein*

Soy protein isolate (SPI), which has a protein content of more than 90%, is the most widely used form of soy protein. Soy protein is isolated from the prepared protein concentrate by removing indigestible polysaccharides, dietary fibers and other components of low refractive index. Flakes subjected to low-humidity heat treatment are extracted in a pH range of 7 to 8 in the solution. The pH of the isolated insoluble filtrate extract solution having insoluble polysaccharide and surplus protein was adjusted to 4.5, which contained a large amount of protein and refined sugar. Through this isolation process, proteins in the bean soaking water are precipitated. When the sugar and trace elements in the filtrate are mixed with the protein, it is difficult to obtain the desired material, so the protein is usually removed by centrifugation or filtration during the process. At the same time, wash and dry the instigated protein into an electrophilic protein (Megías et al. 2016).

In modern food processes, ultrafiltration units are often used to fractionate SPI by molecular weight to grasp the difference between weight of molecules and protein membrane structures, and to study the influence on the physical properties of the membrane and the moisture barrier function of the corresponding SPI fraction (Cho and Rhee 2004).

### 11.3.2 Edible Plastic Based on Modified Soy Protein

The acceptability of the SPI material is more clearly demonstrated by the use of technical and physical properties of auxiliary materials. It is not possible to show appropriate industrial mechanical and physicochemical characteristics, the most feasible method is to mix materials that improve the performance of SPI materials. So far, SPI has been combined to create the optimal compounds with several plasticizers and recyclable polymers (Park et al. 2000). For instance, the engineered plastic glycerol-plastic and critical foil were produced on the basis of the insulate of soy protein and the effects and method dimensions of the formulation system and the application of the reaction methodology was investigated (Nandane and Jain 2018).

1. *Plasticizers*

   Thermomechanical processing is a highly efficient and simple method of preparing edible plastic materials. This method focuses on thermoplastic soy protein isolate (SPI). One of the steps in thermomechanical processing is to add a plasticizer to the SPI matrix to improve the processing performance of the SPI, while suppressing the brittleness of the natural SPI membrane due to reduced hydrogen bonds within the protein chain. The most effective plasticizers are very similar in their polymer structure. Therefore, the compound alcohol composed of a hydroxyl group is often used in the industry as a plasticizer for soy protein isolate (Suderman et al. 2018).

2. *Surfactants*

   Sodium dodecyl sulfate (SDS), which is a good protein dissociation and denaturation agent, is essentially an ionic surfactant. If SDS is mixed in the protein-forming solution, it affects the properties and structure of the cast protein film. Consequently, an effective method for obtaining a brighter and stronger SPI film is to add various types of fatty acids to the film-forming solution such as oleic acid, stearic acid and so on, making the finished product stronger and less likely to shrink (Wu et al. 2018).

3. *Biodegradable polymers*

   In the preparation of bioplastic films, various physical properties of bioplastics can be enhanced by mixing SPI with biodegradable polymers, plasticizers and compatibilizers. Some polymers such as caseinate, propylene glycol alginate (PGA) and polyvinyl alcohol are added during preparation.

   When the SPI and PGA are thoroughly mixed, a covalent complex is formed between the two to further improve the water stability of the film containing the various components. The study found that the addition of PG in the production of biofilm can effectively optimize the film resistance to water and vapour, so SPI-PGA composite film will have potential application prospects in edible packaging field (Rydz et al. 2018). At the same time, the grease resistance of biofilms can be achieved by coating with SPI. During the coating process, specific interactions occur between the cellulose and the SPI to block carbon dioxide and oxygen. By observing the mechanical properties of the SPI-cellulose hybrid film, it was found that the mixed film has good mechanical strength and can be regarded as a good substitute for the non-biodegradable film (Li and Pelton 2005).

   Another material that can be used in the preparation of biomixed membranes has arised from the discovery of caseinate. Despite the problem of poor stretchability and low flexibility, the biomix film with caseinate added was found to have effective features of liquid barrier. It can be denoted that the combination of SPI-caseinate to form a mixed film is an efficient way of enhancing SPI film quality. The transparency of the SPI film can be effectively improved by mixing and mixing sodium caseinate or casein calcium (Fabra et al. 2010).

   Furthermore, the addition of chitin to the SPI has the advantage of enhancing SPI film by improving the tensile strength and Young's modulus. In the production, chitin can be blended with SPI in the presence of a plasticizer such as glycerin by a compression molding mechanism (Zheng et al. 2003).

4. *Lipids and essential oils*

   Wax is generally considered to be a hydrophilic polar material during the selection of food packaging materials and can be used as an important additive for SPI mixing. Since wax is a lipid having a long carbon chain structure, it has good water blocking properties. In the presence of a non-polar solvent, the wax can be isolated from plants and seeds and is not miscible with the aqueous film forming solution to improve the water resistance of the protein film (Kim et al. 2015).

   The hybrid SPI-oil biofilm is one of the effective methods to increase the water vapour barrier property of the mixed membrane. When the essential oil is added to the SPI film, the purpose of preventing microbial growth and lipid oxidation can be achieved. The mechanical properties of the film will vary according to the type of oil. For example, a blend of SPI and cinnamon oil showed high film ductility and elongation resistance (Atarés et al. 2010).

Furthermore, bioplastic films with an essential oil were produced and described with Kefiran-carboxymethyl cellulose. The results showed enhanced tensile strength and contact angle in the main oil concentration but reduced elongation, humidity and water vapour permeability at break time (Hasheminya et al. 2019).

### 11.3.3 Chemical Cross-Linking of Soy Protein

Water boundary capability of soy protein films can increase by calcium cross-linking of SPI. In comparison with the effects of the consequent SPI film, it was shown that the and puncture strength and tensile strength of the calcium-sulphate SPI film were lower than the other when SPI was tested with two forms of calcium salts, namely calcium sulfate and the chloride (Park et al. 2001).

A further technique is to use aldehyde-group chemicals such as glutaraldehyde, glyoxal, formaldehyde and dialdehyde to alter the SPI. It could also build a bridge between the protein chains by responding with the lysine residue amino group where the technical and other features of the crosslinking SPI films have been improved (Vaz et al. 2003). For instance, the transglutaminases are crosslinking enzymes present in protein molecules of distinct origin to catalyse the covalent development of the isopeptide bond. (Sabbah et al. 2019)

### 11.3.4 Radiation Modification

Through the action of γ-radiation, the protein conformation in soy protein changes, and at the same time, protein free radicals are formed, with amino acids undergo oxidation, and protein affected by covalent bond cleavage including reactions of recombination and polymerisation. The irradiated proteins can either be cross-linked or degraded, depending on the radiation dose and nature of the protein. The energy of ultraviolet (UV) compared to ionizing radiation, is considerably weaker. The UV processing can then be used as food wraps for the creation of protein-based films. By using an amine / salt solvent scheme, blend solutions of cellulose and soy protein isolates were developed. Before casting films with a more stable molecular network than non-irradiated alternatives the alternatives were gamma-irradiated. The films contained very little water, regardless their transparency, which indicated the impact of gamma radiation on the chemical structure (Boy et al. 2018).

### 11.3.5 Enzyme Modification

Enzymatic cross-linking to optimize SPI biofilms is considered to be a more advantageous method than chemical cross-linking. The enzyme can configure the polymer into a biopolymer through the covalent cross-linking reaction of the protein, which can achieve the purpose of optimization more safely and efficiently. Therefore, the enzyme cross-linking reaction is more widely used and more popular in the process of improving the SPI film. After the treatment of horseradic peroxidis, the physical and water resistance characteristics of SPI films are found to be improved by catalysing the oxidation of various plastic substrates by hydrogen peroxide. Furthermore, enzymes can selectively connect the structure of the protein matrix between distinct amino acids. Various shapes, activated and disabled, and various preparation conditions, have been investigated to assess the impact on the new bio-plastic film characteristics (Mohammad Zadeh et al. 2018).

### *11.3.6 Surface Modification*

While mixing is a simple and effective technique of making polymeric multi-phase products, interfacial adherence between polymer stages is one of the main draw-backs to overcome when using this technique. Alternatively, laminated films with less compatibility issues can be produced. The full water vapour permeability of biopolymer foils can be minimized by laminating hydrophilic SP films with hydro-phobic covers on the bottom.

Corn zein (CZ) is often used as a film cover since it includes hydrophobic amino acids like proline, leucine and CZ films demonstrated excellent moisture barrier characteristics. Through a technique called thermal compaction, The SPI films could be laminated using corn zein with singular or double coats and the laminate material has been examined with technical and physical properties. The hydropho-bic nature of CZ, compared with the SPI film, strengthened the lamina film's water vapour shield qualities and remained unchanged by its thinner oxygen barrier. Another technique is the preparation of the SPI / PLA multilayer film using a sol-vent casting method that also showed excellent outcomes (Rhim et al. 2007). In recent years, a study shows that comparing the non-modified films with the modi-fied counterpart, the physical structures and water resistance of the modified SPI films were significantly strengthened following the integration of the silane cou-pling agent by unmodified and cellulose nanocrystal modified films. The enhance-ment of thermal stability and water resistance was also taken into account (Zhang et al. 2016).

## 11.4 Economic and Environmental Influence of Biocomposites

In the decades, after biocomposites development due to natural environment and conservation concerns, significant improvements over polymer science have been obtained (Gurunathan et al. 2015). Composite materials are considered biocomposites comprised of one or several biochemical phases. The main goal is to substitute the oil-based product with bio-based materials due to wear and climate change. Food packaging is one of the most significant industries in the globe because it uses more plastic products that influence both human health and the environment (Robertson 2008). In general, plastics is often used as a raw material source of green goods of agricultural residues that are found in nature. Different sections of plants are discovered to be significant source of raw material as fillers for bio composites (Satyanarayana et al. 2009). Natural fibres as organic fillers are found to be extremely important because their price is low, their density per unit volume is high, their strength and use is reduced, as are their sustainable and recyclable properties and they have an incredible effect (Jawaid and Khalil 2011). The specific needs of composites as manufactured materials are met with a vital role for natural fibres and they benefit more than other composite fillers such as low cost, light weight, naturally degradable, environment-friendly, without compromising the inflexibility of materials. Natural fibres have the most enticing impact on the environment because they are $CO_2$-neutral, which means that they don't add excess $CO_2$ to the environment when they are vaporized (Bougueraa et al. 2018).

## 11.5 Further Application of Bioplastics

As bio-composites become of interest, the challenge is to substitute conventional composites with ones that show identical dimensional and technical stability while using clearance, processing and ecological degradation. That is the reason, in future, bio-composite marketing is likely to grow because companies and consumers are conscious of the ecological issues, exploring different implementations and strengthening refining technology efficiency. Strengthened bio-composites fail to acknowledge innovation in developing countries where fibres of this kind are widely accessible (Faruk et al. 2012). A broad spectrum of applications for mass-produced products is intended to use for both shorter and longer-term applications. While biocomposites are feasible and completely re-usable, they can be much more costly if they are entirely organic and biologically-degradable. Although biocomposites have a wide diversity in their properties, the production of natural fibers and their composites may resolve their weaknesses (Gurunathan et al. 2015).

## 11.6 Conclusions

Due to the high degree of toxicity and environmental concerns of petroleum-based polymer products, the design of edible, biodegradable bioplastics gained increased attention. SPI is the by-product of soybean oil production. It is highly biologically compatible, storage capable and biodegradable and is a promising technology in the food industry. Specific modifying techniques including surface modification and bulk modifying have been used to enhance SPI's water resistance and mechanical characteristics, making this an ideal packaging product. In addition, the production, use and removal of fibre strengthened composites that are eco-friendly is seen as a compelling replacement rather than an above-mentioned cut to poly blend materials. These specific characteristics work in the green product setting as a solid foundation for generating new technologies and prospects for biocomposites. As a result, wastewater of legume could provide the chemicals needed (fibres, protein) for cheaper manufacturing and with less need for isolation than legumes themselves.

**Acknowledgments** This book chapter was written thanks to the resources allocated by Lincoln University (New Zealand) to the Bachelor course "FOOD 398 – Research Essay". The author is the Postgraduate Diploma student, Ms. Yanyu Zhang.

## References

Atarés, L., De Jesús, C., Talens, P., & Chiralt, A. (2010). Characterization of SPI-based edible films incorporated with cinnamon or ginger essential oils. *Journal of Food Engineering, 99*(3), 384–391.

Bougueraa, F. Z., El Mouhri, S., & Ettaqi, S. (2018). Experimental analysis of biocomposite Raphia fiber/Chitosan influence of weaving process on mechanical properties. *Procedia Manufacturing, 22*, 180–185.

Boy, R., Bourham, M., & Kotek, R. (2018). Blend films of cellulose and soy protein isolate prepared from gamma irradiated solutions. *European Journal of Engineering and Applied Sciences, 1*(2), 78–83.

Boye, J., Zare, F., & Pletch, A. (2010). Pulse proteins: Processing, characterization, functional properties and applications in food and feed. *Food Research International, 43*(2), 414–431.

Brummer, Y., Kaviani, M., & Tosh, S. M. (2015). Structural and functional characteristics of dietary fibre in beans, lentils, peas and chickpeas. *Food Research International, 67*, 117–125.

Cho, S. Y., & Rhee, C. (2004). Mechanical properties and water vapor permeability of edible films made from fractionated soy proteins with ultrafiltration. *LWT-Food Science and Technology, 37*(8), 833–839.

Cifriadi, A., Panji, T., Wibowo, N. A., & Syamsu, K. (2017). Bioplastic production from cellulose of oil palm empty fruit bunch. In IOP Conference Series: Earth and environmental science (65, 1, p. 012011). IOP Publishing, Bogor, Indonesia.

Dalgetty, D. D., & Baik, B. K. (2003). Isolation and characterization of cotyledon fibers from peas, lentils, and chickpeas. *Cereal Chemistry, 80*(3), 310–315.

Deshpande, A. P., Bhaskar Rao, M., & Lakshmana Rao, C. (2000). Extraction of bamboo fibers and their use as reinforcement in polymeric composites. *Journal of Applied Polymer Science, 76*(1), 83–92.

European Bioplastics. (2016). URL: http://www.european-bioplastics.org/.

Fabra, M. J., Talens, P., & Chiralt, A. (2010). Influence of calcium on tensile, optical and water vapour permeability properties of sodium caseinate edible films. *Journal of Food Engineering, 96*(3), 356–364.

Faruk, O., Bledzki, A. K., Fink, H. P., & Sain, M. (2012). Biocomposites reinforced with natural fibers: 2000–2010. *Progress in Polymer Science, 37*(11), 1552–1596.

Gaspar, A. L. C., & de Góes-Favoni, S. P. (2015). Action of microbial transglutaminase (MTGase) in the modification of food proteins: A review. *Food Chemistry, 171*, 315–322.

Geissdoerfer, M., Savaget, P., Bocken, N. M., & Hultink, E. J. (2017). The circular economy–A new sustainability paradigm? *Journal of Cleaner Production, 143*, 757–768.

Giosafatto, C. V. L., Al-Asmar, A., D'Angelo, A., Roviello, V., Esposito, M., & Mariniello, L. (2018). Preparation and characterization of bioplastics from grass pea flour cast in the presence of microbial transglutaminase. *Coatings, 8*(12), 435.

Gurunathan, T., Mohanty, S., & Nayak, S. K. (2015). A review of the recent developments in biocomposites based on natural fibres and their application perspectives. *Composites Part A: Applied Science and Manufacturing, 77*, 1–25.

Han, J., Shin, S. H., Park, K. M., & Kim, K. M. (2015). Characterization of physical, mechanical, and antioxidant properties of soy protein-based bioplastic films containing carboxymethylcellulose and catechin. *Food Science and Biotechnology, 24*(3), 939–945.

Hasheminya, S. M., Mokarram, R. R., Ghanbarzadeh, B., Hamishekar, H., Kafil, H. S., & Dehghannya, J. (2019). Development and characterization of biocomposite films made from kefiran, carboxymethyl cellulose and Satureja Khuzestanica essential oil. *Food Chemistry, 289*, 443–452.

Huang, S., Liu, Y., Zhang, W., Dale, K. J., Liu, S., Zhu, J., & Serventi, L. (2018). Composition of legume soaking water and emulsifying properties in gluten-free bread. *Food Science and Technology International, 24*(3), 232–241.

Jawaid, M. H. P. S., & Khalil, H. A. (2011). Cellulosic/synthetic fibre reinforced polymer hybrid composites: A review. *Carbohydrate Polymers, 86*(1), 1–18.

Jerez, A., Partal, P., Martínez, I., Gallegos, C., & Guerrero, A. (2007). Protein-based bioplastics: Effect of thermo-mechanical processing. *Rheologica Acta, 46*(5), 711–720.

Kim, H. S., Lee, S. H., Byun, Y., & Park, H. D. (2015). 6-Gingerol reduces Pseudomonas aeruginosa biofilm formation and virulence via quorum sensing inhibition. *Scientific Reports, 5*, 8656.

Korhonen, J., Honkasalo, A., & Seppälä, J. (2018). Circular economy: The concept and its limitations. *Ecological Economics, 143*, 37–46.

Kutoš, T., Golob, T., Kač, M., & Plestenjak, A. (2003). Dietary fibre content of dry and processed beans. *Food Chemistry, 80*(2), 231–235.

Li, X., & Pelton, R. (2005). Enhancing wet cellulose adhesion with proteins. *Industrial & engineering chemistry research, 44*(19), 7398–7404.

Martens, L. G., Nilsen, M. M., & Provan, F. (2017). Pea hull fibre: Novel and sustainable fibre with important health and functional properties. *EC Nutrition, 10*, 139–148.

Megías, C., Cortés-Giraldo, I., Alaiz, M., Vioque, J., & Girón-Calle, J. (2016). Isoflavones in chickpea (Cicer arietinum) protein concentrates. *Journal of Functional Foods, 21*, 186–192.

Mohammad Zadeh, E., O'Keefe, S. F., Kim, Y. T., & Cho, J. H. (2018). Evaluation of enzymatically modified soy protein isolate film forming solution and film at different manufacturing conditions. *Journal of Food Science, 83*(4), 946–955.

Mukherjee, T., & Kao, N. (2011). PLA based biopolymer reinforced with natural fibre: A review. *Journal of Polymers and the Environment, 19*(3), 714.

Nandane, A. S., & Jain, R. K. (2018). Optimization of formulation and process parameters for soy protein-based edible film using response surface methodology. *Journal of Packaging Technology and Research, 2*(3), 203–210.

Nurul Fazita, M. R., Jayaraman, K., Bhattacharyya, D., Mohamad Haafiz, M. K., Saurabh, C. K., Hussin, M. H., & HPS, A. K. (2016). Green composites made of bamboo fabric and poly (lactic) acid for packaging applications—A review. *Materials, 9*(6), 435.

Oomah, B.D., Patras, A., Rawson, A., Singh, N., & Compos-Vega, R. (2011). Chemistry of pulses. In *Pulse foods: Processing, quality and nutraceutical applications* (pp. 9–56), Elsevier, Oxford, UK.

Orenia, R. M., Collado, A., Magno, M. G., & Cancino, L. T. (2018). Fruit and vegetable wastes as potential component of biodegradable plastic. *Asian Journal of Multidisciplinary Studies, 1*(1), 17.

Park, S. K., Hettiarachchy, N. S., & Were, L. (2000). Degradation behavior of soy protein– Wheat gluten films in simulated soil conditions. *Journal of Agricultural and Food Chemistry, 48*(7), 3027–3031.

Park, S. K., Rhee, C. O., Bae, D. H., & Hettiarachchy, N. S. (2001). Mechanical properties and water-vapor permeability of soy-protein films affected by calcium salts and glucono-δ-lactone. *Journal of Agricultural and Food Chemistry, 49*(5), 2308–2312.

Peelman, N., Ragaert, P., De Meulenaer, B., Adons, D., Peeters, R., Cardon, L., Van Impe, F., & Devlieghere, F. (2013). Application of bioplastics for food packaging. *Trends in Food Science & Technology, 32*(2), 128–141.

Puscaselu, R., Gutt, G., & Amariei, S. (2019). Biopolymer-based films enriched with stevia rebaudiana used for the development of edible and soluble packaging. *Coatings, 9*(6), 360.

Rhim, J. W., Lee, J. H., & Ng, P. K. (2007). Mechanical and barrier properties of biodegradable soy protein isolate-based films coated with polylactic acid. *LWT-Food Science and Technology, 40*(2), 232–238.

Robertson, G. (2008). State-of-the-art biobased food packaging materials. In *Environmentally compatible food packaging* (pp. 3–28). Woodhead Publishing.

Rydz, J., Musioł, M., Zawidlak-Węgrzyńska, B., & Sikorska, W. (2018). Present and future of biodegradable polymers for food packaging applications. In *Biopolymers for food design* (pp. 431–467). Academic Press, London, UK.

Sabbah, M., Giosafatto, C.V.L., Esposito, M., Di Pierro, P., Mariniello, L., & Porta, R. (2019). Transglutaminase cross-linked edible films and coatings for food applications. In *Enzymes in food biotechnology* (pp. 369–388). Academic Press, London, UK.

Salmoral, E. M., Gonzalez, M. E., & Mariscal, M. P. (2000). Biodegradable plastic made from bean products. *Industrial Crops and Products, 11*(2–3), 217–225.

Satyanarayana, K. G., Arizaga, G. G., & Wypych, F. (2009). Biodegradable composites based on lignocellulosic fibers—An overview. *Progress in Polymer Science, 34*(9), 982–1021.

Serventi, L., Wang, S., Zhu, J., Liu, S., & Fei, F. (2018). Cooking water of yellow soybeans as emulsifier in gluten-free crackers. *European Food Research and Technology, 244*(12), 2141–2148.

Sood, M., & Dwivedi, G. (2017). Effect of fiber treatment on flexural properties of natural fiber reinforced composites: A review. *Egyptian journal of petroleum, 27*(4), 775–783.

Sreerama, Y. N., Neelam, D. A., Sashikala, V. B., & Pratape, V. M. (2010). Distribution of nutrients and antinutrients in milled fractions of chickpea and horse gram: seed coat phenolics and their distinct modes of enzyme inhibition. *Journal of Agricultural and Food Chemistry, 58*(7), 4322–4330.

Stahel, W. R. (2016). The circular economy. *Nature News, 531*(7595), 435.

Stantiall, S. E., Dale, K. J., Calizo, F. S., & Serventi, L. (2018). Application of pulses cooking water as functional ingredients: The foaming and gelling abilities. *European Food Research and Technology, 244*(1), 97–104.

Suderman, N., Isa, M. I. N., & Sarbon, N. M. (2018). The effect of plasticizers on the functional properties of biodegradable gelatin-based film: A review. *Food Bioscience, 24*, 111–119.

Thakur, S., Chaudhary, J., Sharma, B., Verma, A., Tamulevicius, S., & Thakur, V. K. (2018). Sustainability of bioplastics: Opportunities and challenges. *Current Opinion in Green and Sustainable Chemistry, 13*, 68–75.

Tosh, S. M., & Yada, S. (2010). Dietary fibres in pulse seeds and fractions: Characterization, functional attributes, and applications. *Food Research International, 43*(2), 450–460.

Vaz, C. M., Van Doeveren, P. F. N. M., Reis, R. L., & Cunha, A. M. (2003). Development and design of double-layer co-injection moulded soy protein based drug delivery devices. *Polymer, 44*(19), 5983–5992.

Venkidasamy, B., Selvaraj, D., Nile, A. S., Ramalingam, S., Kai, G., & Nile, S. H. (2019). Indian pulses: A review on nutritional, functional and biochemical properties with future perspectives. *Trends in Food Science & Technology, 88*, 228–242.

Wu, Y., Cai, L., Wang, C., Mei, C., & Shi, S. (2018). Sodium hydroxide-free soy protein isolate-based films crosslinked by pentaerythritol glycidyl ether. *Polymers, 10*(12), 1300.

Yadav, A., Mangaraj, S., Singh, R., Kumar, N., & Simran, A. (2018). Biopolymers as packaging material in food and allied industry. *International Journal of Chemical Studies, 6*, 2411–2418.

Zhang, S., Xia, C., Dong, Y., Yan, Y., Li, J., Shi, S. Q., & Cai, L. (2016). Soy protein isolate-based films reinforced by surface modified cellulose nanocrystal. *Industrial Crops and Products, 80*, 207–213.

Zheng, H., Tan, Z. A., Ran Zhan, Y., & Huang, J. (2003). Morphology and properties of soy protein plastics modified with chitin. *Journal of Applied Polymer Science, 90*(13), 3676–3682.

# Index





L. Serventi, *Upcycling Legume Water: from wastewater to food ingredients*,
https://doi.org/10.1007/978-3-030-42468-8

**X**

Printed in the United States
by Baker & Taylor Publisher Services